Snow Avalanche Hazards and Mitigation
in the United States

Snow Avalanche Hazards and Mitigation in the United States

Panel on Snow Avalanches
Committee on Ground Failure Hazards Mitigation Research
Division of Natural Hazard Mitigation
Commission on Engineering and Technical Systems
National Research Council

NATIONAL ACADEMY PRESS
Washington, D.C. 1990

This study was supported by the Federal Highway Administration under Purchase Order DTFH61-85-P-00311; U.S. Department of the Interior/Office of Surface Mining, Contract No. J5130137; U.S. Geological Survey, Grant No. 14-08-0001-G1101; Federal Emergency Management Agency, Contract No. EMW-85-K-2202; U.S. Department of the Army; Naval Facilities Engineering Command; and National Science Foundation, Master Agreement No. 82-05616 to the National Academy of Sciences. Any opinions, findings, and conclusions or recommendations expressed in this report are those of the committee and do not necessarily reflect the views of the sponsoring agencies.

Library of Congress Catalog Card No. 90-62814

Limited number of copies available without charge from:

Committee on Ground Failure Hazards Mitigation Research
Division of Natural Hazard Mitigation, HA 286
2101 Constitution Avenue, N.W.
Washington, DC 20418

Additional copies of this report are available from:

National Academy Press
2101 Constitution Avenue, NW
Washington, DC 20418

S218

Printed in the United States of America

Frontispiece photos: (top) Dry snow (powder) avalanche from slab release triggered from helicopter (courtesy of W. Porton); (bottom) Wet snow avalanche in a developed area (courtesy of FISAR).

COMMITTEE ON GROUND FAILURE HAZARDS MITIGATION RESEARCH

DAVID B. PRIOR (*Chair*), Louisiana State University, Baton Rouge
GENEVIEVE ATWOOD, Utah Geological and Mineral Survey, Salt Lake City
DAVID S. BROOKSHIRE, University of Wyoming, Laramie
RHEA L. GRAHAM, Consultant, Placitas, New Mexico
A. G. KEENE, Department of Public Works, Los Angeles, California
F. BEACH LEIGHTON, Leighton & Associates, Inc., Irvine, California
GEORGE G. MADER, William Spangle & Associates, Portola Valley, California
H. CRANE MILLER, Attorney, Washington, D.C.
WILLIAM J. PETAK, University of Southern California, Los Angeles
DWIGHT A. SANGREY, Rensselaer Polytechnic Institute, Troy, New York
ROBERT L. SCHUSTER, U.S. Geological Survey, Denver, Colorado
JAMES E. SLOSSON, Slosson & Associates, Van Nuys, California
DONALD R. SNETHEN, Oklahoma State University, Stillwater
DOUGLAS N. SWANSTON, Forestry Sciences Lab, Juneau, Alaska
BARRY VOIGHT, Pennsylvania State University, University Park

Liaison Representatives

N. B. BENNETT III, Division of Geology, Bureau of Reclamation, Denver, Colorado
LEON L. BERATAN (retired), Office of Nuclear Regulatory Research, Nuclear Regulatory Commission, Washington, D.C.
C. Y. CHEN, (formerly) Office of Surface Mining, Washington, D.C.
ROBERT A. CUMMINGS, Society of Mining Engineers, Engineers International, Inc., Tucson, Arizona
DONALD G. FOHS, Construction Maintenance and Environmental Design Division, Federal Highway Administration, McLean, Virginia
ROBERT P. HARTLEY, Solid and Hazardous Waste Research Division, U.S. Environmental Protection Agency, Cincinnati, Ohio
DARRELL G. HERD, U.S. Geological Survey, Reston, Virginia
THOMAS L. HOLZER, Geological Society of America, U.S. Geological Survey, Menlo Park, California
BENJAMIN I. KELLY, U.S. Army Corps of Engineers, Washington, D.C.
LOUIS KIRKALDIE, Soil Conservation Service, Washington, D.C.
PAUL KRUMPE, Office of Foreign Disaster Assistance, Agency for International Development, Washington, D.C.
ADRIAN PELZNER, Forest Service, U.S. Department of Agriculture, Washington, D.C.
J. E. SABADELL, Division of Biological and Critical Systems, National Science Foundation, Washington, D.C.
CHI-SHING WANG, Division of Health and Safety Technology, Bureau of Mines, Washington, D.C.
DAVIS E. WHITE, Manufactured Housing and Construction Standards Division, U.S. Department of Housing and Urban Development, Washington, D.C.
MICHAEL YACHNIS (retired), Facilities Engineering Command, U.S. Department of the Navy, Alexandria, Virginia
ARTHUR J. ZEIZEL, Office of Natural and Technological Hazards, Federal Emergency Management Agency, Washington, D.C.

Staff

Riley M. Chung, Director
Abe Bernstein, former Senior Staff Officer
Barbara Bodling, Editor
Jennifer T. Estep, Administrative Secretary
Patricia T. Green, Research Aide
Susan R. McCutchen, Administrative Assistant
Shirley J. Whitley, Project Assistant

PANEL ON SNOW AVALANCHES

BARRY VOIGHT (*Chair*), Pennsylvania State University, University Park
B. R. ARMSTRONG, Fulcrum Inc., Denver, Colorado
R. L. ARMSTRONG, University of Colorado, Boulder
D. BACHMAN, Consultant, Crested Butte, Colorado
D. BOWLES, Montana State University, Bozeman, and Utah Department of Transportation, Alta
R. L. BROWN, Montana State University, Bozeman
R. D. FAISANT, Attorney, Palo Alto, California
S. A. FERGUSON, Northwest Avalanche Center, Seattle
J. A. FREDSTON, Alaska Mountain Safety Center, Anchorage
J. L. KENNEDY, Attorney, Sun Valley, Idaho
J. KIUSALAAS, Pennsylvania State University, University Park
E. R. LACHAPELLE, University of Washington, Seattle
R. C. McFARLANE, Camrose Lutheran College, Camrose, British Columbia
R. NEWCOMB, American Avalanche Institute, Wilson, Wyoming
R. PENNIMAN, Consultant, Tahoe City, California
R. PERLA, NHRI Environment, Canmore, Alberta

Acknowledgments

Principal authors of the report are panel members B. Voight (**Editor**), B. R. Armstrong, R. L. Armstrong, D. Bowles, R. L. Brown, S. A. Ferguson, J. A. Fredston, J. Kiusalaas, R.C. McFarlane, and R. Penniman.

Valuable contributions to report planning were made by

D. Bachman, Consultant, Crested Butte, Colorado
R. D. Faisant, Attorney, Palo Alto, California
D. Fesler, Consultant, Alaska Mountain Safety Center, Anchorage
J. L. Kennedy, Attorney, Sun Valley, Idaho
E. R. LaChapelle, University of Washington, Seattle
M. Martinelli, Jr. (retired), U.S. Forest Service
R. Newcomb, American Avalanche Institute, Wilson, Wyoming
R. Perla, NHRI Environment, Canmore, Alberta

Written or oral contributions were provided by D. Abromeit, J. Anderson, G. Borrel, H. Boyne, D. S. Brookshire, S. Burns, E. Burr, R. Christiansen, S. Colbeck, J. E. Fagan, R. Feuchter, G. Fiebiger, L. Fitzgerald, D. Foley, G. L. Freer, C. Fuchs, D. Gallagher, S. K. Gerdes, W. Good, H. Gubler, J. M. Herbert, L. Heywood, J. Hoagland, C. Jaccard, R. J. Janda, A. Judson, R. Kattelmann, L. Knazovicky, W. Kockelman, T. E. Lang, J. O. Larsen, J. C. Leiva, P. Lev, K. Lied, G. Mace, N. Maeno, R. A. Mandahl, R. T. Marriott, D. M. McClung, A. I. Mears, J. Montagne, M. Moore, T. Nakamura, K. Niemczyk, H. Norem, R. B. Olshansky, J. C. Paine, P. Schaerer, R. A. Schmidt, C. Stethem, J. M. Taillandier, T. W. Tesche, C. Tolton, F. Valla, K. F. Voitkovskiy, W. Walters, E. Wengi, O. Wieringa, C. Wilbour, K. Williams, N. Wilson, and C. Wuilloud.

The editor apologizes for any omissions, which are probably inevitable although unintentional.

Contents

Executive Summary

Snow avalanche is a type of slope failure that can occur whenever snow is deposited on slopes steeper than about 20 to 30 degrees. Avalanche-prone areas can be delineated with some accuracy, since under normal circumstances avalanches tend to run down the same paths year after year, although exceptional weather conditions can produce avalanches that overrun normal path boundaries or create new paths. Unlike other forms of slope failure, snow avalanches can build and be triggered many times in a given winter season.

In the United States, as elsewhere, snow avalanches are a mounting threat as development and recreation increase in mountain areas: the recorded incidence of avalanches is greater, and the number of people affected by avalanche events and avalanche hazard is also increasing. Data from avalanche accidents show that avalanche activity occurs in about one-third of the states and is a significant hazard in much of the West, where avalanches are the most frequently occurring lethal form of mass movement. Present annual mortality due to snow avalanches exceeds the average mortality due to earthquakes as well as the average mortality due to all other forms of slope failure combined. Avalanches pose hazards that affect a significant sector of the public; involve a number of private organizations; and require cooperation and action by government agencies at the federal, state, and local levels. Avalanche hazard causes economic loss to residents, businesses, transportation systems, and government agencies. It can have a negative impact on the local economy of many mountain regions and affects the management of federal lands. Avalanche-related litigation is a growing problem.

Hazard mitigation requires measures ranging from appropriate land-use management and effective building codes in avalanche-prone areas to the timely issuance of emergency warnings and programs of public education. Centralized avalanche information and forecast centers currently play an important hazard-reduction role in Colorado, Utah, and Washington. These centers are funded by a variety of state, federal, and private organizations, but the funding base is not secure in all cases and their survival is an issue of concern. The Alaska Avalanche Center lost its state funding after the winter of 1985–1986 and has not resumed operations.

Direct avalanche control is appropriate for areas used intensively by the public, though

1

it is too difficult or costly for the vast areas open to recreational use. Control is ordinarily exercised through structural engineering systems or by the artificial release of built-up snow cover. Engineering techniques such as snowsheds and wedges can be applied to modify terrain so as to divert moving snow from facilities, and various fence structures have been devised to stabilize snow on mountainsides. Artificial release techniques focus on the frequent release of small avalanches to inhibit the formation of a large avalanche and employ explosive charges delivered by hand, artillery, or mechanical conveyance. Improved standards and operational procedures need to be instituted for the safe deployment of explosive systems. This problem should be addressed at the federal level since U.S. military weapons and stockpiles are involved. Issues to be considered include safety training, certification standards, the inventory of critical ammunition, spare parts, aging ammunition, ammunition storage and transportation, and the growing problem of lost shells. A conservative calculation suggests that several thousand armed but unexploded military artillery shells deployed for avalanche control now exist in backcountry areas of the United States. Since many operational problems associated with artillery control are eliminated by cable delivery systems, further attention to cable delivery should be encouraged.

Despite the destructive nature of snow avalanches and the hazards they pose to mountain residents and vacationers, the United States lacks coordinated national leadership on avalanche issues. There is currently no national program for avalanche prediction, land-use planning, research, and education. There is an inadequate basis for the exchange of information among government personnel, scientists, engineers, forecasters, and control specialists. Support for avalanche research has almost vanished in the United States, although research is relevant to all aspects of avalanche control and hazard mitigation. Closely associated with the decline of research is a reduction of the national institutional capability in avalanche expertise and a decline in technology transfer that seeks to take advantage of the extensive avalanche work done in other countries.

From the late 1930s until 1985, the U.S. Department of Agriculture/Forest Service played the major role in snow avalanche mitigation in the United States. Toward this end, the agency conducted its own research, funded outside research, provided technology transfer and guidance, and set policy in areas of avalanche safety and education. Due in part to restricted funding, the Forest Service abnegated this responsibility in 1985. Although avalanche hazard continues and is increasing, individuals involved in the identification, evaluation, and solution of problems related to avalanche hazard no longer have a specific agency or facility to consult for guidance and expertise.

There is much that can be done to reduce avalanche hazards in the United States. There are obvious needs for geologic and engineering research, for the development of hazard-delineation techniques, for improved understanding of avalanche initiation and the dynamic processes that influence structural controls, for expansion of forecasting services and better-coordinated dissemination of information about avalanche hazards, and for the resolution of serious problems associated with the use of explosives. There are no widely accepted guidelines or regulatory approaches for taking avalanche hazards into account in community planning, and the programs that exist vary considerably. Apart from lands under federal jurisdiction, the reduction of avalanche losses through land-use management and the application of effective building codes are essentially functions of local government, with enabling legislation by the state. Avalanche insurance, although in principle a viable option, is virtually unobtainable.

Reduction of avalanche hazards should be viewed as a national goal requiring national leadership. Such leadership is essential for promoting more effective implementation of existing organizational capabilities and improving cooperative support, information, and technical assistance. The federal government should assume its specific but limited responsibilities for avalanche hazard delineation and control, including the development of relevant methodologies on a variety of scales, pilot mapping and control demonstrations, and avalanche mapping and control in support of the missions of federal agencies. Research under national leadership should be undertaken to improve the technical base for avalanche forecasting, control, land-use planning, and public warning systems through (a) interdisciplinary research by appropriate federal agencies and (b) support and maintenance of a research capability by universities through funding by the National Science Foundation.

To assist the federal government in assuming a more active and sustained role, the panel recommends the formation of a short-lived interagency task force or committee to initiate program coordination among federal agencies having responsibilities related to slope failure, snow research, forecasting centers, and the administration of federal lands containing avalanche-prone areas. Next, sustained nationwide coordination of avalanche management and research programs could be performed most effectively by a national-level committee composed of representatives from government, academia, industry, and professional organizations. Whatever its nature, there should be adequate representation of the specific interests of federal, state, and local agencies and of private groups with responsibilities for various aspects of avalanche mitigation. The purpose of the committee would be to provide direction and momentum for the solution of these problems. Such a committee could be organized and maintained over the long term by a committee of the National Research Council (NRC) charged with reduction of natural hazards or, alternatively, a panel within the Committee on Glaciology of the NRC's Polar Research Board.

In the development of any national program, useful ideas can be obtained from the successful and cost-effective national avalanche-mitigation programs in operation in Japan, France, Norway, the U.S.S.R., and Switzerland, where avalanches have long been recognized as the single greatest natural hazard to winter activities in mountain areas. The International Decade for Natural Disaster Reduction program (National Research Council, 1987) suggests possible avenues for international cooperation in this area and should provide motivation toward the establishment of an effective avalanche-mitigation capability in the United States.

1
Snow Avalanche Problems

This report by the Committee on Ground Failure Hazards Mitigation Research addresses the problems and mitigation issues concerning snow avalanche hazards in the United States. Other reports by the committee have considered problems due to landslides (National Research Council, 1985) and ground subsidence (National Research Council, in press). The present report is the first publication on snow avalanches by any National Research Council committee; therefore, it is essential to include both a general and historical perspective in order to provide sufficient background for discussion of current problems. This information is not available elsewhere through any single published source.

The purpose of the report is to provide national, regional, and local governments; government agencies; and private decision makers with an overview of the snow avalanche situation in the United States and to outline steps that can be taken to minimize domestic avalanche problems. Four major points are emphasized:

1.   Support for avalanche programs has diminished alarmingly at a time when increasing numbers of people are using mountain areas for recreation and commercial and other types of development are increasing in formerly remote areas.

2.   The incidence of avalanche accidents is increasing and is expected to continue to increase in the future.

3.   There is a lack of nationwide coordination, accepted standards, and effective information flow among those involved in avalanche mitigation.

4.   There are no standardized procedures for avalanche control and equipment testing. Control techniques and equipment that use explosives have specific hazards and problems that must be addressed.

Snow avalanches have caused natural disasters as long as mountainous areas have been inhabited. They are a common occurrence in mountainous terrain throughout the world, wherever snow is deposited on slopes steeper than about 20 to 30 degrees. In the United States, where avalanches are the most frequent form of lethal mass movement, avalanche hazard exists from the lower-elevation coastal mountain ranges to the higher mountains of the continental interior.

5

By definition a snow avalanche is simply snow moving rapidly down sloping terrain. A moving avalanche may also contain soil, rock, vegetation, or water, but by definition the initial failure that triggers an avalanche occurs within the snowpack or at the interface between snow and subjacent terrain. Avalanches range from a harmless trickle of loose snow descending to a new angle of repose to a huge and devastating mass of snow moving at high speed down a long steep slope, with enough energy to destroy everything in its path. It is important to know that, unlike other ground-failure hazards such as rockslides, which once released are spent, snow avalanches automatically "reload" with each snowfall and can "fire" several times in a given year.

Small avalanches or sluffs run in uncounted numbers each winter, while larger avalanches, which may encompass slopes several kilometers wide and include millions of tons of snow, release infrequently but have the potential to inflict the greatest destruction. Avalanches of moderate size can damage structures and have the ability to bury, injure, and kill people. In the United States approximately 10,000 avalanches are reported each winter, with an estimated 10 to 100 times that number occurring unobserved or unreported (Armstrong and Williams, 1986).

Terrain and weather patterns combine to determine the frequency of avalanche events. Large frequent snowstorms in combination with steep slopes will produce a high number of avalanches during a given winter season. Under ordinary circumstances, avalanches tend to run in the same location and down the same paths year after year, with the danger zones often becoming well known. However, exceptional weather conditions can produce avalanches that overrun their normal path boundaries or even create new paths where none existed for centuries (Fitzharris, 1981), as illustrated by the destruction in Switzerland of a 573-year-old stone building in 1957 (Friedl, 1974). Unusually high snowfall can provide short-lived but great hazard, in which even historically stable slopes may become dangerous (Figure 1).

A factor in most avalanche releases is the presence of structural weaknesses, often induced by internal changes in snow cover. Hence, a large overburden of snow alone may not result in avalanching if it is internally strong and anchored to the layer below, but a shallow snow layer can slide from a mountainside if the snow is poorly bonded to the underlying material. Snow avalanches represent a complex problem in mechanical stability; thus, attempts to provide a better understanding of the phenomenon have focused primarily on the physical processes taking place within the constantly changing winter snow cover and the dependence of those processes on temperature and other meteorological factors.

A hazard arises whenever property or human activity lies in the path of a potential avalanche. Snow avalanche hazard has been familiar to inhabitants of the European Alps and Scandinavia for many centuries, but it is a more recent problem in the United States. During the active period of gold and silver mining from 1880 to 1920, approximately 400 people were killed by avalanches in Colorado, many trapped within structures. More recently the primary hazard has been to individuals engaged in recreation activities, with deaths and injuries frequently occurring at some distance from developed facilities. Such events have the potential to affect the local economy of many mountain regions and to exert a significant effect on federally managed lands.

U.S. citizens may also be endangered by avalanche hazards abroad. Those exposed to risk include not only Alpine recreationists (Vila, 1987) but also military personnel, as illustrated by the 1986 NATO exercise in Norway, during which 31 men were struck by a naturally released avalanche; 16 were killed and 11 injured (Kristensen, 1986). In a distinct

FIGURE 1   (a) An "attractive" potential development site in a century-old lodgepole pine forest on a fan beneath Deadman Gulch, Colorado Front Range, 1976. Small avalanches over previous several decades had been contained by adjacent gullies. (b) The same area in May 1984, showing the effect of a "100-year" dry snow avalanche. This avalanche far exceeded the boundaries of previously recorded events and destroyed many acres of the pine forest that had colonized in the runout zone for over a century. These photographs provide valuable before-and-after documentation of the "design avalanche," the event magnitude that should be considered in land-use planning and design of exposed facilities. Because most avalanche paths have not recently produced an event of design magnitude, many planners and others tend to ignore or underestimate the potential avalanche threat. (Courtesy of A. Mears and Paula J. Lehr)

category are the military catastrophes of the Tyrol in World War I, where estimates of avalanche-caused fatalities ranged from 40,000 to 80,000 (Fraser, 1966).

Avalanche danger is alleviated in three fundamental ways: by modifying the terrain, by modifying the snow cover, and by modifying human behavior. A number of engineering techniques have been used to divert or deflect moving snow from facilities; other techniques are used to prevent destructive avalanches from releasing. Reforestation provides a natural form of protection, but avalanche risk may substantially increase in the near future due to forests dying or deteriorating as a result of air pollution.

The most common technique for reducing avalanche hazard is to artificially release potential avalanches at a selected safe time. This practice inhibits the formation of large avalanches by producing more frequent smaller ones. While this method may not provide as high a degree of protection as some terrain-modification techniques, it is less expensive in the short term; the technique is commonly used at ski areas and along highways and railroads. Avalanches are usually released by explosive charges, detonated on or near the snow surface close to the expected fracture point. Such charges are placed by hand or delivered to the slope using some form of artillery or mechanical conveyance.

Because avalanches can affect winter vacationers, widespread public education about avalanches is of particular importance. Instruction on how to evaluate and avoid avalanche-prone terrain and on rescue techniques is important for reducing hazards to downhill and cross-country skiers and snowmobilers. The highly mobile nature of these activities makes control with structures and explosives difficult. Centralized avalanche information and forecast centers such as those located in Colorado, Utah, Washington, and some other areas are an essential ingredient in avalanche education. In some cases land-use management and zoning can be used to protect the public in avalanche-threatened areas.

Yet despite the increasing hazards posed by snow avalanches to mountain residents and tourists in the United States, there is no coordinated national program for avalanche mitigation. There is no recognized national leadership, no systematic means to improve understanding of avalanche processes or to improve mitigation procedures, and no adequate and comprehensive mechanism for information transfer and exchange.

2
The Avalanche Phenomenon

AVALANCHES—A TYPE OF GROUND FAILURE

The geotechnical community recognizes five ways that slope failure can occur under the force of gravity (Varnes, 1978; Ground Failure, 1985). Material may fall freely (or almost freely) through the air—from a cliff, for example. It may topple or tilt over a pivot point. It may slide downward along an identifiable surface or narrow zone that is curved or spoon shaped (rotational slide or slump) or relatively planar (translation slide). It may spread laterally across a slope or flow as a thick fluid—sometimes very rapidly, sometimes so slowly as to be barely perceptible (creep). Slope movements involving two or more of these types of movements are termed *complex*.

As both the kind of material involved and the motions that occur are of importance in slope failure investigations, these factors are commonly used to classify slope movements. The approach can be extended to include snow and ice, and the resulting classification places the full range of gravitational movements of snow and ice within the logical format familiar to engineers (Table 1). This emphasizes the important point that snow avalanches are an integral part of the U.S. landslide problem (Voight and Ferguson, 1988; Voight, 1978).

Ice avalanches usually begin with the slow basal sliding and creep of ice caps and glaciers that overhang cliffs. Instability produces true falling or toppling, followed by partial disintegration and flowage. Ice avalanches can be devastating, as in Switzerland in 1965 when 88 workers engaged in dam construction were killed by an avalanche released by the Allalin glacier (Fraser, 1966; Mellor, 1978; Roethlisberger, 1978). In the United States, large ice avalanches are known to happen in Alaska (Slingerland and Voight, 1979) and in the Cascade Range (Williams and Armstrong, 1984a; Voight, 1980, 1981; Voight et al., 1981; Williams, 1934; Bleuer, 1989).

Falls and sometimes topples characterize the failure of snow cornices and the release of snow from building roofs (Paine and Bruch, 1986; Taylor, 1985), but typically the initial failure mechanism of the snow cover is translational sliding, utilizing a sloping surface of weakness within the snow cover or at the ground-snow interface. Continuation of movement

TABLE 1 Classification of Slope Movements in Snow, Rock, and Soil Based on Kind of Material and Type of Movement (Modified from Varnes, 1978)

TYPE OF MOVEMENT			Bedrock	Engineering Soils		Snow
				Predominantly coarse	Predominantly fine	
Falls			Rock fall	Debris fall	Earth fall	Snow fall
Topples			Rock topple	Debris topple	Earth topple	Snow topple
Slides	Rotational		Rock slump	Debris slump	Earth slump	Snow slump
	Translational	Few Units	Rock block slide	Debris block slide	Earth block slide	Snow block slide
		Many Units	Rock Slide	Debris slide	Earth slide	Snow slide
Lateral Spreads			Rock spread	Debris spread	Earth spread	Snow spread
Flows			Rock flow	Debris flow	Earth flow	Snow flow
			(rock creep)	(soil creep)		(snow creep)
Complex			Combination of two or more principal types of movement			

leads to breakdown of individual snow slabs and, if sufficient disintegration occurs, to rapid mass flowage.

Like rock avalanches, most snow avalanches are complex phenomena involving several basic types of motion in succeeding phases. Use of the standard geotechnical terminology for precisely depicting this complexity can lead to an unwieldy vocabulary (rockfall-rapid rock fragment flow, for example). As a result, other classification schemes have been developed over the years, and these have been widely adopted by snow scientists (e.g., Table 2; UNESCO, 1971). In one such scheme, two basic types of snow avalanches are recognized—point release and slab—based on the conditions at the release zone or place of origin. Each scheme is subdivided according to whether the snow is dry, damp, or wet; whether the movement originates within the snow layers or involves the entire snow cover down to the ground surface; and whether the motion is mainly over ground, through the air, or mixed (Perla, 1978a, 1980).

A point release or loose snow avalanche (sluff) is the result of a small amount of cohesionless snow slipping out of place, moving downslope, and encountering additional cohesionless snow, such that the failure progresses and spreads out into a characteristic inverted V-shaped pattern. Point releases usually occur either within the cohesionless near-surface layers of newly fallen snow or within the wet surface snow resulting from melt

TABLE 2 Morphological Classification of Snow Avalanches (after UNESCO, 1971)

Zone	Criterion	Alternative characteristics, Denominations, and Code		
Zone of origin	A Manner of starting	A1 Starting from a point (loose snow avalanche)	A2 Starting from a line (slab avalanche)	
			A3 Soft A4 Hard	
	B Position of sliding surface	B1 Within snow cover (surface layer avalanche)	B4 On the ground (full-depth avalanche)	
		B2 (New snow B3 (Old snow fracture) fracture)		
	C Liquid water in snow	C1 Absent (dry snow avalanche)	C2 Present (wet-snow avalanche)	
Zone of transition (free and retarded flow)	D Form of path	D1 Path on open slope (unconfined avalanche)	D2 Path in gulley or channel (channeled avalanche)	
	E Form of movement	E1 Snow dust cloud (powder avalanche)	E2 Flowing along the ground (flow avalanche)	
Zone of deposition	F Surface roughness of deposit	F1 Coarse (coarse deposit) F2 Angular F3 Rounded blocks clods	F4 Fine (fine deposit)	
	G Liquid water in snow debris at time of deposition	G1 Absent (dry avalanche deposit)	G2 Present (wet avalanche deposit)	
	H Contamination of deposit	H1 No apparent contamination (clean avalanche)	H2 Contamination present (contaminated avalanche)	
			H3 Rock debris, soil H4 Branches trees	
			H5 Debris of structures	

conditions. Point releases usually involve small volumes of snow and can be predicted without much difficulty, so they generally present only a small degree of hazard.

In contrast, slab avalanches initiated within cohesive snow cover on slopes steeper than 25 degrees provide most of the avalanche hazards and are the primary focus of defense and control measures. Failures occur when the shear load parallel to the slope exceeds the shear strength of supporting layers. In this case the layer of cohesive snow, poorly anchored underneath, fractures as a continuous single unit. Given relatively homogeneous snow properties, the fracture may propagate for a great distance across a slope and may incorporate a large volume of snow into the moving avalanche. Fractures may extend as much as several meters into the snow cover. Prediction of slab avalanches is difficult

TABLE 3 Scale of Snow and Ice Avalanches

ORDER-OF-MAGNITUDE ESTIMATES

Size	Potential effects	Vertical descent (m)	Volume (m^3)	Impact pressure (Pa)	(psi)[a]
Sluffs	Harmless	10	$1\text{-}10^3$	$<10^3$	<0.15
Small	Could bury, injure, or kill a human	$10\text{-}10^2$	$10\text{-}10^2$	10^3	0.15
Medium	Could destroy a wood frame house or auto	10^2	$10^3\text{-}10^4$	10^4	1.5
Large	Could destroy a village or forest	10^3	$10^5\text{-}10^6$	10^5	15
Extreme	Could gouge landscape, world's largest avalanches (Himalayas, Andes, Alaska)	$10^3\text{-}5\times10^3$	$10^7\text{-}10^8$ (includes ice, soil, rock, mud)	$10^5\text{-}10^6$	15-150

[a]Pa is approximately 1.5×10^{-4} psi.

SOURCE: After McClung and Schaerer (1981).

because the location of initial failure is frequently well below the surface, within layers that accumulated weeks or months earlier (LaChapelle, 1985; McClung, 1979; Armstrong, 1979; Ferguson, 1984a,b; Perla, 1978a). These layers are hard to locate and monitor prior to actual failure. Slab avalanches present a significant hazard due to this difficulty of prediction, in addition to their potential for release over large areas. Escape from these avalanches can be difficult or impossible (see frontispiece). The hazard to activity and structures in the avalanche runout zone is high due to the large volumes of snow that can be mobilized by a slab release. Table 3 provides a qualitative scale of the destructive potential of snow avalanches and related physical parameters.

Wet snow avalanches present additional problems due to their high mobility and erratic style of runout (see frontispiece; Martinelli, 1984). Also important is the rapid mass movement of water-saturated snow known as a slush avalanche or slushflow. Analogous to the mudflows and debris flows of conventional geotechnology, slushflows are major natural hazards in Scandinavia, the U.S.S.R., Canada, Greenland, and Alaska (Hestnes, 1985; Hestnes and Sandersen, 1987; Onesti, 1985, 1987; Nobles, 1965; Nyberg, 1985; Rapp, 1960).

The multiple-hazard concept can also be important in relation to avalanches, such as downstream flooding due to breakout of avalanche-dammed rivers (Fraser, 1966; Williams,

1934) and water-wave damage or flooding due to large avalanches into water bodies and mine-tailings impoundments (Slingerland and Voight, 1979; Vila, 1987; NGI, 1984, 1986).

CAUSES OF AVALANCHE RELEASE

Use of the term natural release implies the occurrence of a trigger beyond human control. In a simplified sense, natural releases fall into two end-member categories: in one case the load increases while the strength remains generally unchanged (e.g., rapid loading by snowfall); in the other the load remains approximately constant while the strength decreases (e.g., strength loss due to melting). Of course, hybrid situations also exist in which both load and strength vary over time. Artificial release usually results from the placement of explosives by hand, military weapons, or specialized avalanche control equipment. Skiers, snowmobilers, and climbers crossing an avalanche starting zone may also cause artificial release.

Point release avalanches occur when cohesionless snow rests on a slope that is steeper than its natural angle of repose. Failure is localized, and the mechanism is not difficult to understand. In contrast, slab release occurs when a cohesive cover of snow rests above a layer of lesser strength along which the eventual sliding failure occurs, when shear stress exceeds shear resistance. Slab release typically results from a complex series of events, often originating within a snow cover creeping downslope (McClung, 1987). When differential stresses cause localized failure, load is transferred to the adjacent snow structure; if this additional load cannot be sustained, cracks are initiated and propagated by a rapid increase in stress due to stress concentration and the release of stored strain energy. The failure process can also be initiated by a tree or rock acting as the anchor and source of concentrated stress to a slowly sagging snow cover or by additional dynamic loading from a falling object.

In terms of predicting slab avalanche occurrence, the average mechanical values are frequently less important than the range of values (Gubler, 1988; Sommerfeld and King, 1979; Ferguson, 1984b; Conway and Abrahamson, 1988). For example, overall conditions can be moderately stable, while localized stress concentrations in an otherwise strong and homogeneous snow cover allow local slab failures to expand and incorporate a major portion of a slope's snow cover.

Failure is most common during or soon after a heavy snowstorm, when potentially weak layers cannot strengthen rapidly enough through crystal-to-crystal bond formation (sintering) to support the increasing shear load of new snow. In addition, weakness may originate as the snow recrystallizes by temperature-sensitive metamorphism deep within the snow pack (Colbeck, 1980, 1987; Perla and Ommanney, 1985; R. L. Armstrong, 1977, 1981, 1985; Marbouty, 1980). Metamorphism within snow is a continuous process that begins when snow is deposited and continues until it melts. The processes causing changes in crystallinity are complex, but it is known that important roles are played by mass transfer, water vapor diffusion, and temperature and temperature gradient (Colbeck, 1982, 1987; Sommerfeld, 1983; Gubler, 1985). The susceptibility of the snow cover to rapid changes (over hours or days) in layer and bond strength is a reflection of the proximity of the ambient temperature of the snowpack to the melting temperature of ice. Mechanical and thermal properties and conditions for snow are closely intertwined, and the interplay of geomechanics, thermodynamics, and meteorology contributes to the complexity of stability analysis.

Avalanches can also be triggered by direct dynamic loads due to falling cornices, the passage of skiers through the starting zone, rockfalls, or elastic waves from blasting or earthquakes (Hackman, 1965; LaChapelle, 1968; Voight and Pariseau, 1978; Johnson, 1978, 1980; Brown, 1980). Finally, rain-on-snow events may cause wet snow avalanches and slushflows. While rain-induced slides constitute a small proportion of all avalanches, such events can produce substantial damage (Hestnes, 1985; Hestnes and Sandersen, 1987; Kattelmann, 1984, 1987; Moskalev, 1966; Onesti, 1987; Ambach and Howorka, 1965).

GEOGRAPHIC DISTRIBUTION OF AVALANCHE HAZARD

Figure 2 shows the severity of avalanche hazard in the United States; the assignment of severity classes is qualitative but is based on avalanche fatality data for the winters of 1950–1951 through 1987–1988 (cf. Armstrong and Williams, 1986). These data indicate that avalanche activity occurs in nearly all western states (Washington, Oregon, California, Idaho, Nevada, Montana, Utah, Wyoming, Colorado, New Mexico, and Alaska); in Minnesota; and in the northeastern states of Maine (McFarlane, 1986), New Hampshire, and New York. Thus, about one-third of the 50 states face avalanche hazard during winter months. In at least two states, avalanches kill more people than any other natural hazard. In addition, avalanches can occur in man-made snow (Avalanche Review, 1988), and fatalities, injuries, and damage—as well as litigation—have been produced by avalanches from sloping roofs (Taylor, 1985; Paine and Bruch, 1986; Nakamura et al., 1981).

Current annual U.S. mortality due to snow avalanches exceeds the average number of deaths from earthquakes and generally exceeds the combined average number of deaths due to all other forms of landslides [about 12 per year according to Jahns (1978) for the period 1925–1975; accurate statistics are not available for landslides, cf. Schuster and Fleming (1986)]. Single avalanche events killed 96 people in Washington in 1910, 70 in Alaska in 1898, and 40 in Utah in 1939 (Gallagher, 1967; Perla, 1970). U.S. avalanche fatalities were routinely reported to the U.S. Forest Service from about 1960 to 1984 and since then to the Colorado Avalanche Information Center (Gallagher, 1967; Williams, 1975; Williams and Armstrong, 1984a). Alaska leads the nation in the number of reported avalanche accidents per capita; over 3,000 Alaskan avalanche events involving humans have been documented. Between 1980 and 1985, Alaska recorded 441 events affecting people; in those events 278 persons are known to have been trapped, injured, or killed (J. A. Fredston, Alaska Mountain Safety Center, written communication, 1986). However, apart from fatalities, accident data for other states are less complete, and probably fewer than 10 percent of nonfatal avalanche accidents are reported (Armstrong and Williams, 1986). Currently about 140 Americans are reported each year to be caught in avalanches, 65 being buried and 17 killed (Armstrong and Williams, 1986).

Enough private property is threatened by avalanches to have prompted the enactment of local avalanche zoning ordinances in California, Colorado, Idaho, Utah, and Washington (Armstrong and Williams, 1986; Clark, 1988; Niemczyk, 1984). Zoning regulations have not yet been adopted elsewhere, although hazards have been recognized. In the last 105 years, for example, over 80 structures within a 10-mile radius of Juneau, Alaska, have been hit or destroyed by avalanches, and several large avalanche paths from Mt. Juneau—which towers above the city—threaten residential housing (Hart, 1972; LaChapelle, 1972, 1981; Hackett and Santeford, 1980). One of these avalanche paths, the North Behrends Avenue path, contains 40 houses, a motel, a highway, a large boat harbor, and a school in its runout zone. Since 1890, avalanches released from its 43-acre starting zone have run down this path

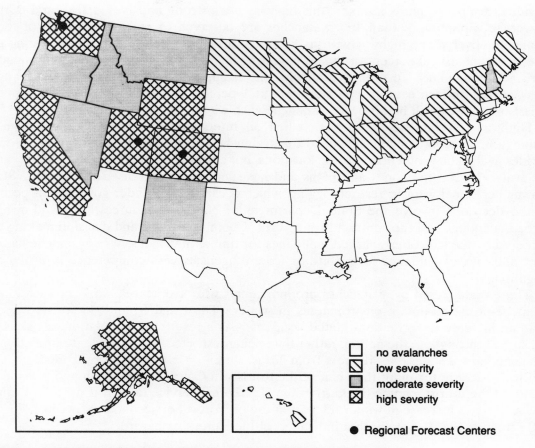

no avalanches
low severity
moderate severity
high severity

● Regional Forecast Centers

FIGURE 2   Qualitative indication of the severity of snow avalanches by state.

almost to tidewater at least five times, and in 1962 an avalanche physically relocated two dozen homes and took roofs off many others. The Behrends Avenue path has a significant potential for a large-scale avalanche disaster, but despite several detailed studies identifying specific hazard areas within city limits (Hart, 1972; LaChapelle, 1972, 1981), construction in known runout zones continues (J. A. Fredston, Alaska Mountain Safety Center, personal communication, 1986).

Frequent avalanche activity threatens transportation corridors along numerous year-round highways and railroads in such areas as Washington state; the Alaskan coastal region; California's Sierra Nevada; and canyons in Utah, near Jackson Hole, Wyoming, central and western Colorado, and western Montana. For example, in the decade preceding 1986, 205 avalanche events in Alaska blocked highway traffic, with 30 vehicles hit or disabled; 274 events blocked railroad traffic, with 21 cars derailed; and 2 aircraft were damaged (D. Fesler, Alaska Mountain Safety Center, personal communication, 1986). Eleven state and federal highways in Colorado are also susceptible to avalanches, and during the winter of 1983–1984, Colorado highways were closed by natural avalanches on 60 days (Williams and Armstrong, 1984b).

One example of a high-risk highway is U.S. Highway 550 in southwestern Colorado, where avalanches threaten a third of the road from Ouray to Coal Bank Pass. Ninety-three individual avalanche paths intersect this highway (Armstrong and Ives, 1976), and during an average winter more than 100 avalanches are observed to reach it. If each of these avalanches covered the highway with its historical maximum quantity of debris during the same storm, nearly 30 percent of the 36-mile-long highway would be covered by avalanche debris. Although such an event is unlikely, this example provides a measure of the high potential risk. In fact, extensive data indicate a 79 percent probability that on this highway at least one vehicle will be hit by an avalanche each winter (B. Armstrong, 1980).

Highway travelers have also been killed on Interstate 90 in the Cascade Range in Washington, creating a significant need for avalanche control. Here, dense forest growth provides avalanche protection at some locations, but clear-cut timber harvesting has created new avalanche starting zones and paths and has significantly increased the hazard. State highway personnel want to restrict logging, which is carried out under government permit, but the decisions involve the State Department of Natural Resources, the land-owning railroad company, and the contract logging firm. Communications and decision making are hampered by the lack of established guidelines for timber management in avalanche hazard zones and lack of guidance from state or federal agencies with competence in avalanche mitigation.

Those endangered by avalanches are individuals who live, travel, work, or vacation in avalanche-prone mountain environments in winter. The trend from 1940 to the present shows an increase in recreation-related accidents. As a result, the population at risk is, in fact, spread throughout the nation, rather than being restricted to the resident population of avalanche hazard states. Vacationers from Texas, Illinois, New York, Florida, Georgia, and California account for more than half of the tourists in Colorado and Utah and provide 90 percent of the tourist income. Data from the U.S. Forest Service Region II office in Denver show a 15 times increase in winter recreation use on Forest Service lands between 1970 and 1980 (Trogert, 1981)—yet another indication of the increasing hazard.

ECONOMIC COSTS OF U.S. AVALANCHES

Avalanches damage and destroy public, commercial, and private property and forest lands and result in costs for restoration, maintenance, and post de facto litigation. No comprehensive study has been attempted of the economic impact of snow avalanches in the United States.

Direct costs can be defined as the cost of maintenance, restoration, or replacement due to damage of property or structures within the boundaries of a specific avalanche. All other costs from avalanches are indirect and include (1) reduced real estate values in areas threatened by avalanches, (2) loss of productivity of forest lands, (3) loss of industrial productivity as a result of damage to land or facilities or interruption of services, (4) loss of tax revenues on properties devalued as a result of avalanches, (5) cost of measures to mitigate additional land or facility damage, (6) loss of access to recreation lands and facilities, (7) cost of lost human productivity due to injury and death, and (8) the cost of litigation as a consequence of avalanches. Some of these indirect costs are difficult to measure and tend to be ignored. As a result, most estimates of avalanche costs are far too conservative. If rigorously determined, indirect costs probably exceed direct costs.

Direct and indirect costs can be further subdivided into ex-anti and ex-post costs (D. S. Brookshire, University of Wyoming, unpublished manuscript, 1986). Ex-anti costs are those incurred prior to an avalanche event for preventative measures such as forecasting and control; added research and design in planning; and the cost of materials, labor, and delay time in the implementation phase. Ex-post costs are incurred following an avalanche event and include the costs of search and rescue personnel and equipment, reconstruction, loss of productivity, property damage, injury, and loss of life.

It is useful to further divide direct and indirect costs by categorizing them as either public (government) or private. Direct public costs have traditionally been limited to the ex-anti cost of avalanche forecast and information centers, hazard mitigation in the form of structural and/or active control measures for roads or public work facilities, and the ex-post cost of clearing debris and repairing damage. Historically, the federal government funded ongoing avalanche research, an ex-anti cost, but in 1985 it withdrew all research support.

Direct private costs (both ex-anti and ex-post) have been borne primarily by utility companies and the winter recreation industry in the form of structural and/or active mitigation measures, forecasting, and facility repair and maintenance. Private property owners have mainly incurred ex-post costs with a major avalanche potentially resulting in financial ruin for affected individuals because of the unavailability of avalanche insurance or other means of spreading the costs of damage. With the recent implementation of avalanche hazard zoning ordinances by some county and municipal governments, ex-anti costs in the form of mapping and additional planning and construction, as well as indirect economic losses from property devaluation, are also beginning to be incurred by private home and property owners.

It should also be noted that for major avalanche events the costs are sustained at all levels of the public and private sectors for emergency services, disaster aid, and reconstruction.

EXAMPLES OF ECONOMIC IMPACT

The economic impact of snow avalanches can be determined only through accurate and complete cost data, but in most cases such data are either nonexistent or are hidden in the cost data of other programs. In a limited number of instances, accurate and specific cost data are available that illustrate the type and extent of the economic impact of avalanches in the United States. In the following examples, cost figures have not been adjusted to current dollars.

The costs associated with rescue depend on the magnitude of an avalanche, but a recent incident adjacent to the Breckenridge ski area in Colorado provides a fairly typical example. The avalanche occurred on February 18, 1987, on U.S. Forest Service (USFS) land just outside the boundary of the Breckenridge ski area; four people were killed. The ski area's rescue costs were $35,000, including helicopter time, salaries, and miscellaneous costs; the cost to the Colorado Search and Rescue Board was approximately $39,250. In addition, the USFS incurred undisclosed losses for in-house and town meetings and for preparation of anticipated litigation.

Litigation is a major economic cost. An avalanche at the Alpine Meadows ski area in California on March 31, 1982, resulted in seven deaths and caused extensive damage to the base area's facilities. Property damage to buildings, vehicles, and equipment was approximately $1.5 million, not including rescue costs and timber loss. However, the families of a number of the victims subsequently sued Alpine Meadows, the USFS, Placer County, Southern Pacific Land Company, and the food service company operating the lodge's cafeteria on the day of the accident (Penniman, 1986, 1987). Several suits were settled out of court for undisclosed amounts. In an ensuing court battle involving three plaintiffs, the jury exonerated Alpine Meadows and rendered a verdict of nonnegligence for standards of "ordinary care" (Garbolino, 1986; Gerdes, 1988). In addition to undisclosed amounts for out-of-court settlement from the subsequent appeal and a suit against the USFS, litigation costs for the defense, not including the USFS's defense costs, were approximately $700,000, and litigation costs for the plaintiff were about $800,000. The out-of-court settlement cost potentially could have been $14 million (J. Fagan, University of San Francisco, personal communication, 1986). [The theory of strict liability as it applies to avalanche explosive control is discussed by Fagan and Cortum (1986). Liability of adjoining property owners and liability for failure to maintain unstable land, among other matters, are examined by Olshansky and Rogers (1987).]

Other types of losses result from the impact on property values of ordinances enacted by local municipalities in the mountain states. These ordinances either prohibit construction in high and moderate avalanche hazard areas or call for special engineering practices. An example is provided by Placer and Nevada counties, California, where avalanche ordinances were enacted in 1982 as a direct result of the Alpine Meadows avalanche. A real property loss of $712,000 was estimated in this case, reflecting the difference in the value of property sold before and after enactment of the ordinance (R. Penniman, consultant, Tahoe City, California, personal communication, 1986). Here, the restrictions for building in high and moderate avalanche hazard zones are so severe that no lots (out of a total of 52) have been sold there. Thus, by applying the average percentage loss of real property values to these no longer salable properties, an additional loss to property value can be estimated at $7,100,000. Yet the loss extends beyond the potential sale of property: when property values decrease, tax revenues also are affected. The potential yearly tax loss in this case is $72,000, based on percentages of 1.0103 and 1.1003 quoted by the Placer County Assessor's Office for Squaw Valley and Alpine Meadows, respectively.

Hundreds of mountain communities in the United States have the potential for avalanche incidents similar to those experienced at Breckenridge and Alpine Meadows. An incident could involve large numbers of people within or near ski areas and on roads and residential and commercial areas in surrounding communities. A single avalanche disaster at any one of these areas is likely to trigger action by litigants and communities similar to those listed above and have similar far-reaching economic effects.

The overall costs associated with avalanche control, mitigation, and damage are difficult to estimate. Annual loss figures from the Washington State Department of Transportation, for example, amount to approximately $330,000 (M. Moore, Northwest Avalanche Center, Seattle, Washington, personal communication, 1986). However, this figure does not include salary costs for personnel employed directly in avalanche control nor the costs of plowing, snow removal, or avalanche control on Cayuse Pass, Chinook Pass, and Washington Pass, which are normally closed during the winter season but require considerable labor and equipment to clear avalanche debris prior to closing and upon reopening (Wilbour, 1986; cf. Sherretz and Loehr, 1983).

Similarly, the state's annual loss figure does not account for adverse impacts on ski resorts, restaurants, lodging, or other businesses in the area. That these impacts are extensive can be inferred from a report by R. Milbrodt, city manager of South Lake Tahoe, which is affected by the periodic closure of U.S. Highway 50 for avalanche control:

> If an avalanche results in road closure (U.S. 50) there is an immediate impact if full closure occurs from lost visitor days. Should the closure only be temporary (less than a day) it is unlikely that economic impact will be noticed in the short-term. In the long-term, however, our market research

suggests that fear of possible closure from avalanche control does discourage visitation. This indirect impact is not measurable, but can be significant.

There are also impacts beyond the scope of our immediate community. For example, during the 1983 event we found that Carson City, Nevada and Placerville, California business activity suffered declines. Those communities previously were of the opinion that they were not dependent on South Lake Tahoe for economic activity and much surprise arose from the losses during U.S. 50 closure. However, we simply do not have any measurements for all of these economic impacts. (R. Milbrodt, city manager, South Lake Tahoe, written communication)

Similarly, serious local economic losses due to highway and rail closures have been cited for the Glacier National Park area in Montana (Butler, 1986). It is clear from such examples that economic losses from avalanches, though usually unspecified as to amount, are of considerable local significance.

A conservative estimate of $11.4 million was placed on the costs incurred from Alaskan avalanches for the period 1977–1986 (D. Fesler, Alaska Mountain Safety Center, Anchorage, personal communication, 1986). This estimate reflects the actual costs of known damage resulting from known avalanche events affecting bridges, buildings, vehicles, power lines, railroads, highways, and miscellaneous structures. Again, however, it does not account for snow removal from highways and railroads or for the cost of avalanche rescues or commercial losses due to delays.

This small sample of losses incurred as a result of avalanches in the United States clearly illustrates that costs recur each winter and will continue to recur year after year. With high media visibility, avalanches have some potential to influence the multibillion-dollar tourism industry that sustains the local economies of many mountain regions. In Colorado, for example, the ski industry accounts for 25 percent of employment, 21 percent of personal income, and 45 percent of housing construction in the western slope region (Frick, 1985). Yet public and private policies in the United States have focused on short-term solutions to the avalanche problem, using relatively inexpensive mitigation measures. Inevitably this effort is subject to local failure, with resulting loss of life and property and a lingering economic impact.

Other nations have learned that such losses cost far more in the long term than do structural mitigation measures that can render avalanches harmless. In Canada, for example, the cost of avalanche research is about 10 to 15 percent of the profits to management, a figure that fully justifies research expenditures (D. M. McClung, National Research Council of Canada, written communication, 1989). An international perspective on hazard management thus is important and is provided in succeeding chapters. First, however, the U.S. policy toward avalanche management is outlined.

3
Avalanche Management Policy in the United States

HISTORICAL DEVELOPMENT

Not until the westward expansion did avalanches become a hazard in the United States. Mormon settlers moving into Utah, railroad workers laying track across California's Sierra Nevada, and prospectors and miners exploring the mountains of the Rockies all quickly learned of the dangers of snow avalanches. On a single day in 1898 on the Chilkoot Trail in Alaska, 70 gold rushers were killed (J. Fredston, Alaska Mountain Safety Center, personal communication, 1986).

Avalanches became a fact of life in the late nineteenth century mining communities. Inhabitants were forced to recognize avalanche hazards and consider legal measures to protect themselves and their property. After a devastating avalanche destroyed the newly built Sampson Mine buildings in southwestern Colorado, killing one man, the local newspaper suggested that expert advice be sought when locating buildings in potential avalanche terrain. Another avalanche, which destroyed the 13-year-old buildings of the nearby Highland Mary Mine, prompted the following (B. Armstrong, 1976):

> Again, buildings should not be put up where there is . . . danger of slides, and we believe that the Colorado legislature should pass a law making it a penal offense for mining superintendents who have buildings put up in dangerous places or where there is the possibility of a slide sweeping them away. Until such a law is passed, there will be lots of chances taken in the erections of buildings (January 27, 1887, San Juan newspaper).

This was one of the earliest public calls for government to enact avalanche hazard legislation.

Following the disastrous 1905–1906 winter in Colorado's San Juan Mountains, with dozens of fatalities and extensive property loss due to avalanches, the editor of the Silverton Standard proposed a full-scale zoning plan for the area, with three different types of protective controls: the power to issue or withhold building permits or licenses based on the location of a building, the gathering of statistics on avalanche location and frequency, and the actual forecasting of avalanche events so that buildings could be evacuated. These goals would eventually be met—but not for more than half a century (B. Armstrong, 1976).

In the twentieth century, mining declined throughout the west and the population in mountain communities was much reduced. As a result, avalanche threats to life and property also decreased. Not until the late 1930s was there a resurgence of concern about avalanche danger. The impetus was the development of downhill or alpine skiing. Faced with the fact that the recreational ski areas being developed were on U.S. Forest Service (USFS) land and that avalanches posed a hazard, the USFS took action. Before the ski lifts were built at the Alta area in Utah, the USFS established the country's first avalanche study center and assigned C. D. Wadsworth as a snow ranger (Kalatowski, 1988). This was the beginning of federal involvement in the avalanche problem.

Development of instruments and techniques for avalanche management began in 1946 (LaChapelle, 1962). In an effort to address the growing number of avalanche accidents, the National Ski Patrol System (NSPS), in 1949, sponsored a visit by the Swiss avalanche expert André Roch. Roch investigated avalanche sites and trained USFS rangers, highway workers, and ski patrollers in snowcraft and avalanche management. He was the first to identify the complexity of U.S. avalanche problems due to the different snow climates of the maritime, intermountain, and central regions (Roch, 1949; cf. LaChapelle, 1966; Armstrong and Armstrong, 1987; Mears, 1984).

Following Roch's visit, the USFS took the lead role in avalanche forecasting, control, research, rescue, and education in the United States, and by 1955 it had established avalanche centers at Berthoud Pass, Colorado; Alta, Utah; and Stevens Pass, Washington. These were operational centers whose purpose was to monitor avalanches in their different climatic areas, supervise control work, and administer efforts to learn and teach more about avalanches.

The Alta Avalanche Study Center took the lead role, guiding the activities of the other two centers and establishing pioneering experiments on explosive control measures (Atwater, 1968; LaChapelle, 1962; Kalatowski, 1988; Hoagland, 1988; see also M. M. Atwater Collection, University of Oregon Library). The center also established the first training programs in avalanche forecasting and control as part of the USFS's snow ranger training, so that avalanche problems in ski areas and along highways on USFS land could be resolved by local snow rangers. In 1971 this training was formalized into the USFS National Avalanche School.

In the 1960s the national focus for avalanche problems began to shift from Alta to the USFS Rocky Mountain Forest and Range Experiment Station in Fort Collins, Colorado. In 1961 this station provided assistance for Colorado State University to invite Hans Frutiger, another Swiss avalanche specialist, to spend a year as a guest researcher (Frutiger, 1964). After Frutiger's visit, USFS research on snow and avalanches accelerated, and federal funds were made available for avalanche research. By the mid-1960s the research initiated and carried out at the Alta Avalanche Study Center under the auspices of administrative studies was transferred to the research branch of the USFS and assigned to the Alpine Snow and Avalanche Project of the Fort Collins Station (U.S. Department of Agriculture/Forest Service, 1971). The long-term records of mountain weather and avalanche occurrence, begun at Alta, Stevens Pass, and Berthoud Pass, were continued, and additional reporting sites were established throughout the western United States.

Alpine snow and avalanche research continued at Fort Collins throughout the 1970s and early 1980s, and a broad range of snow and avalanche problems were investigated. Data were collected on avalanche accidents, avalanche frequency, and mountain weather throughout the western United States, a regional avalanche forecast center was established, a three-phase National Avalanche School was developed, and an international exchange program was set up with the Swiss Federal Institute for Snow and Avalanche Research. The Alpine Snow and Avalanche Project also funded university research and produced technical and lay publications. Avalanche bulletins for Colorado were issued by the USFS regional forecast center in the early 1970s, and subsequently the USFS helped establish other regional avalanche forecast centers in Utah, Washington, and Alaska to provide daily public forecasts of backcountry avalanche conditions on USFS lands.

An internal USFS document reviewed the program in 1973 and reported the following (Martinelli, 1973):

> In the United States, the Forest Service program has displayed professional leadership for the past 3 decades. . . . The need for continued research and the opportunities to apply the findings to operational problems are enough to justify the existing program. . . . If the Forest Service relinquishes its leadership, avalanche work in this country will probably dwindle to a series of unrelated, short-term studies centered at two or three university research groups that are highly dependent on government grants. This is likely to result in a decline in snow safety.

CURRENT STATUS

After 1981 the USFS made a conscious effort to move away from its role in avalanche affairs. In 1982 it helped establish the National Avalanche Foundation, a private nonprofit foundation, to aid the transition of responsibility for snow avalanches from the USFS to other agencies. At present, the foundation is controlled by representatives of USFS recreational management, the ski industry, and the NSPS. Until 1987 its primary function was to administer the National Avalanche School, a task previously undertaken by the USFS. In 1987 the NSPS assumed responsibility for this school.

In 1985 the USFS terminated the Alpine Snow and Avalanche Project at Fort Collins, thus ending its funding and direct involvement in avalanche research. No other government agency has assumed this role. The USFS has also reduced or relinquished its involvement in avalanche work with some regional centers.

After the USFS relinquished administration, the Alaska Avalanche Center was funded by the State of Alaska and administered by the University of Alaska's Arctic Environmental Information Data Center (Hackett and Fesler, 1980). However, the center lost its funding after the 1985–1986 winter because of the state's economic problems and has not resumed operations. The Colorado Avalanche Information Center, which forecasts for all areas of the state, continues to receive some financial support from the USFS; however, it is administered by the Colorado State Department of Natural Resources and must rely on a broad group of organizations for essential financial support. The Utah Avalanche Forecast Center, serving the northern Wasatch Range, is solely supported by the USFS. The Northwest Avalanche Center, forecasting for the Cascades in Washington and Oregon and the Olympic Mountains in Washington, is administered by the USFS with financial support from the National Park Service, the Utah State Park Service, the Northwest Ski Area Association, and the Washington State Highway Department. The National Weather Service provides housing and cooperates with all the centers.

USFS snow rangers in California, Montana, Wyoming, and Idaho provide some services similar to those of the larger regional centers, but their forecast areas are small and involve less use. In these states daily information about snow and avalanche conditions in avalanche-prone backcountry areas is issued as part of the snow rangers' other duties.

Because avalanches occur or originate in large part on lands administered by the USFS, the National Park Service, or the Bureau of Land Management, some responsibility for avalanche mitigation falls on the federal government to assure protection of the general public and private enterprises on federal lands. A similar argument supports the involvement of the U.S. Geological Survey and the USFS in volcano hazards (Bailey et al., 1983; Brown, 1982). At many locations, such as Mono County, California, avalanches that threaten private property originate on federal lands (S. Burns, Planning Director, Mono County, written communication, 1987; M. Martinelli, Jr., U.S. Forest Service, written communication, 1989).

The justification for USFS involvement in avalanche problems is in part related to winter backcountry use, since federal lands—especially national forests—are designed to provide opportunities for unconfined outdoor recreation. USFS policy is to "enhance recreation experiences through a minimum of regulation and law enforcement" (USFS Manual 2303, Item 7) and "regulate users only to the extent necessary for user safety" (USFS Manual 2350, 3, Item 5). A 1987 review of existing policies carried out by Colorado's White River National Forest, prompted by 11 avalanche-caused deaths in the state during the winter of 1986–1987, stated that the USFS should review its level of financial support for the Colorado Avalanche Information Center to ensure that the USFS is providing its fair share, since such centers provide "a very valuable service to users of the National Forests" (Woodrow, 1986).

The federal government retains specific though limited responsibilities, as defined by Public Law 93-28, the Disaster Relief Act of 1974. This law authorizes federal agencies to be prepared to issue disaster warnings to state and local officials (Sec. 202) and to provide technical assistance to states in developing preparedness plans and programs, including hazard reduction, avoidance, and mitigation (Sec. 201) for "any . . . landslide, mudslide, snowstorm . . . or other catastrophe in any part of the United States" (U.S. Department of Agriculture/Forest Service, 1983, pp. 340-348). Although it is clear that snow avalanches may be included under this umbrella listing of natural hazards, ambiguity exists for purposes of response as to whether to group avalanches under the category of landslide, snowstorm, or other. Such ambiguity may contribute to the present lack of federal agency involvement.

Policy on avalanche matters has generally been lacking at the state level. Few states have enacted legislation that applies unambiguously to avalanche mitigation. In 1973 the State of Washington enacted the Land Development Act, which requires the disclosure of any natural hazard on or around a development. This law applies only to developments of 10 or more lots, with smaller ones exempt from the requirement. This legislation is a direct result of the Yodelin avalanche accident, in which 7 cabins were damaged and 13 people were buried, 4 of whom were injured and 4 killed. The residents of the Yodelin development sued the State of Washington, the developer of the Yodelin homesites, and the real estate agency that represented the developer. The appellate court decision acknowledged that the state could be tried for negligence if it could be shown that it had assumed the common law duty to warn the appellants and had either done it improperly or had not done it at all, thus acknowledging in limited fashion the state's "duty to warn" (*Brown* v. *MacPherson's, Inc.*, 1975; Gerdes, 1988).

In 1974 the Colorado Legislature passed House Bill 1041, which made avalanches a matter of state concern and required individual counties to consider assessment of natural hazards for land-use decisions. As a result of this legislation and with financial support from the state, many of the mountainous counties in Colorado now have some type of natural hazards plan that includes a specific avalanche hazard section (Ives and Plam, 1980; Mears, 1980).

Further, as part of the responsibility of the Colorado Geological Survey under House Bill 1041 (C.R.S. 1973, 24-65.1-101, et seq.), and when the need seemed most urgent, snow avalanche hazards were identified at several areas where development was contemplated (Mears, 1976, 1979).

In Utah, where avalanches have resulted in more fatalities than has any other natural hazard, the Geologic Hazards Information Act 1984 HB-28 specifically identified snow avalanches as a significant hazard to public safety and property (UGMS, 1983). The legislation required that hazard maps be prepared and made available to the public. Utah's governor, S. M. Matheson, has expressed his commitment to the concept of disclosure of known hazards to potential property buyers (UGMS, 1983).

On the local or municipal level, avalanche hazard policy is highly variable. In Ketchum, Idaho, the municipal government passed an avalanche zoning ordinance that pays particular attention to the "duty to warn" by providing that the public be notified of avalanche potential within all designated avalanche areas, as determined by detailed studies (Mears, 1980). Ordinances and restrictions for development are currently under study in Mono County, California (S. Burns, Planning Director, Mono County, written communication, 1987). In Alaska municipal governments continue to ignore studies and recommendations for avalanche zoning. In the 1950s a proposed school site in Juneau was relocated because of avalanche hazard, in response to an effort involving the U.S. Geological Survey (Twenhofel et al., 1949; LaChapelle, 1972). But faced with a precise definition of this serious hazard (Hart, 1972; LaChapelle, 1972; Frutiger, 1972; Hackett and Santeford, 1980), Juneau has for almost 20 years refused to enact an avalanche zoning ordinance. In May 1985 the Anchorage Assembly voted down a proposal to establish avalanche hazard zones (Armstrong and Williams, 1986). The law would have identified potentially dangerous avalanche areas on maps assembled by avalanche experts, required landowners to notify prospective buyers or lessors of the hazard, restricted development in the zones, and imposed strict building standards.

COMMENTS

1. Throughout U.S. history, government policy toward avalanche hazard has been one of laissez-faire. In most cases, policy was formulated only when individual government agencies were directly involved and a policy was required. Public policy evolved in response to problems—a reactive approach that tackles each problem on an individual basis, rather than establishing broad national policies. The USFS's involvement with avalanches, for example, came about as a result of avalanche problems at ski-area developments on national forest lands.

2. Currently, there is no national management of avalanche information, research, forecasting, zoning, or education. Nor is there any formal coordination of avalanche-related activities at other levels of government. The USFS has retreated from avalanche hazard management by withdrawing its financial support for education, research, and general management of its centralized repository for avalanche data and information. Policies for avalanche hazard zoning exist mainly at the local level, among municipalities and counties. Only a few states have land-use policies that specifically refer to avalanches.

3. The problem of fragmentation is repeated with avalanche forecasting. Many agencies are involved in forecasting avalanches—the USFS, National Park Service, state de-

partments of natural resources, state highway departments, and local public and private organizations—yet no unifying policy exists and financial support is insecure. Similarly, the administration of avalanche education and forecasting programs is fragmented by state, region, and agency.

4. The states' role should lie between that of the federal government and the local government: to determine priorities, guide efforts, and coordinate statewide the results of federal, university, and private work. If its role as middleman is done well, the state can free the local governments through enabling legislation for the work they are best suited to do (see, e.g., Jochim et al., 1988, for an excellent example of a state response to landslide mitigation).

5. The private sector has been slow to take responsibility, partly because of the high costs and federal restrictions on established avalanche control procedures. State and federal agencies have been equally slow to assume leadership in avalanche mitigation, since many avalanche accidents occur on USFS lands. It should be noted that research and development to defray some of the rising costs of avalanche mitigation is not being supported by any organization.

6. Yet in regard to the Disaster Relief Act of 1974, legitimate arguments can be summoned to include snow avalanches under the aegis of snowstorms or landslides, which after all is merely a popular term to encompass the variety of styles of slope failure in a variety of materials (Varnes, 1978; Voight, 1978). Landslides are accepted as a serious national problem, and although federal funding has been insufficient to allow full compliance with the responsibilities indicated by the Disaster Relief Act of 1974, the need for a national landslide hazard reduction effort is recognized (National Research Council, 1985). The U.S. Geological Survey's landslide research program, which has responsibility for important parts of this effort (U.S. Geological Survey, 1977, 1981, 1982), had peak funding of about $4 million in the 1980s (including Geologic and Water Resources Division activities; G. Wieczorek, U.S. Geological Survey, personal communication, 1988). In comparison, funding for (now-defunct) snow avalanche research by the USFS amounted to about $250,000 during the peak year.

When comparing snow avalanche and landslide programs, it should be recognized that many types of slope failure exist, not all of which are hazardous to life or cause severe economic loss. Instead, different kinds may prevail in different regions, at different times, or under different climatological conditions. One cannot therefore speak of a national problem involving rock avalanches, debris avalanches, or debris flows individually. It is only when all these examples are grouped together that the cumulative severity of the slope failure problem can be appreciated. In this respect, snow avalanches appear at only slight disadvantage when compared to the cumulative effects of all other types of slope failure. Snow avalanches kill about 17 persons each year, compared to perhaps 12 on average for all other types of slope failure and about 25 in peak years (Jahns, 1978; Schuster and Fleming, 1986).

With some federal attention to the avalanche problem warranted, and with snow avalanches recognized as a type of slope failure, a possibility that should be further explored is the incorporation of some snow avalanche process and hazard-delineation research into the U.S. Geological Survey's slope failure program. Such a linkage could be accomplished to the mutual benefit of both the national snow avalanche hazard-mitigation effort and the U.S. Geological Survey's capability to exercise leadership in slope failure research. No single federal agency carries out research in slope stability, the responsibility being shared by the U.S. Geological Survey, the USFS, the Agricultural Research Service, the Bureau of Reclamation, and the U.S. Army Corps of Engineers, among others. Similarly, it should not be assumed that one agency must necessarily carry out all federal responsibilities for snow avalanches.

4
An International Perspective on Avalanche Management

Avalanche hazards are a global management concern, affecting Austria, Bulgaria, Canada, Chile, China, Czechoslovakia, France, Greenland, Iceland, India, Italy, Japan, Nepal, New Zealand, Norway, Pakistan, Peru, Poland, Romania, Scotland, Sweden, Switzerland, the United States, West Germany, and the U.S.S.R. Each nation manages its avalanche hazard in relation to its form of government, appropriate historical precedents, political influences, social concerns, the economic climate, and its technological sophistication. These variables affect the emphasis placed on avalanche management, the implementation strategies adopted, and the relative success or failure of the strategies in achieving individual program objectives.

When devising strategies to reduce avalanche impacts, the parameters of snow dynamics are targeted as well as the distribution and activities of people. The physical aspects of avalanche control are discussed in Chapter 5; the human element is of concern in this chapter. Of particular interest are the strategies that have been implemented frequently or that have shown greater sophistication in the past decade: these include avalanche legislation and regulation, avalanche zoning, and hazard mapping insurance or disaster relief (Kockelman, 1986). These strategies are examined in this chapter in an international context, with the following countries selected for comparison: Austria, Canada, France, New Zealand, Norway, Switzerland, the United States, and the U.S.S.R. This sample reflects both the dominant avalanche management interests and a diverse geographic distribution.

LEGISLATION

Switzerland has both significant avalanche hazard and a well-documented legislative mandate (Frutiger, 1980). The Swiss Federal Confederation oversees 26 cantons, 12 of which are mountainous. The confederation does not have the power to legislate avalanche safety, but it does have the authority to direct the cantons and communes to legislate. Cantons are responsible for enacting planning or zoning and building laws, and communes are responsible for the safety of life and property and, as such, can impose land-use controls directly.

In 1951 the confederation passed an act requiring cantons and communes to adopt avalanche hazard zoning plans. The response was poor; so in 1972 the confederation again passed an act ordering cantons to zone avalanche hazards, threatening severe consequences for lack of action. Most cantons responded: six have zoning codes expressly for avalanches, five have zoning codes for hazards in general, and one still has no zoning (Frutiger, 1980). Swiss avalanche zoning is carried out with the aid of subdivision regulations and/or regulations and codes that require the subdivider to submit maps showing avalanche hazards. Maps are subsequently reviewed by local governments and adjusted if necessary. Building regulations and codes specify the conditions that must be met for development to be permitted in or near an avalanche path, including structural design, construction materials, and even avalanche defenses. Thus, both subdivision and building regulations outline the conditions under which a structure can be built. Construction can begin only after zoning maps indicate that building at a particular location will not jeopardize lives and property.

In addition, numerous legal precedents serve to define the law if regulations are absent. Of particular significance are decisions concerning safe access and the duty to warn involving the sale of property (Frutiger, 1980). In the latter case, property sales can be contested if avalanche hazard becomes evident later; a seller can be charged with intentional fraud if he or she does not inform a prospective buyer of a known hazard.

The French Alps have avalanche situations similar to those in Switzerland, but land speculation in the 1950s compounded the problems (de Crecy, 1980). Regulation is centralized, with the central government having the obligation to define avalanche hazard zones. In 1980 the French government created an interministry committee to study mountain safety, which subsequently recommended that the task of mapping avalanches be assigned to the Institute Géographique Nationale (Cazabat, 1972). Once compiled, avalanche maps became the property of the Minister of Agriculture. Under the French Code de l'Urbanisme (town planning code), areas with avalanche hazards are delimited by a 1974 prefect decree, and within these zones construction can be restricted. The hazard zones must be defined before a building permit can be issued, and the government controls development in these zones through a building code that requires certain structural specifications. Code limitations can include restrictions on the density of buildings and mandatory evacuation or seasonal use of buildings during periods of high avalanche hazard (de Crecy, 1980). As of 1979, approximately 50 mountain communities in France had avalanche zoning plans representing approximately one-half of the communities with avalanche-prone terrain.

The main innovation since 1980 has been the Plan d'Exposition aux Risques (PER), a risk map with legal connotations defined in a 1982 law bearing on natural hazard insurance (Brugnot, 1987). Although 1982 was a year of decentralization laws in France, the government decided that responsibility for natural hazards would not be transferred to individual communities, as was responsibility for most other urban planning problems. A subsequent 1984 decree specified the procedures for producing a PER. The PER is developed under the authority of the prefect (as government representative), but approbation of the communities is required. If agreement is lacking between local councils and the prefect, final decisions regarding PER are rendered by a national conciliation court.

Municipal boards regulate development in Norway. The national building code determines whether residential land use will be permitted, stating that the "ground can only be built on if there is sufficient safety against subsidence, inundation, landslides, etc." (Hestnes and Lied, 1980). If there is inadequate avalanche protection, areas must be classified in area development plans as dangerous (Ramsli, 1974). Zoning per se is not a mandatory requirement of municipal governments. The alternative approach of avalanche mapping became the responsibility of the Norwegian Geotechnical Institute in 1972; previously, mapping was carried out by the University of Oslo's Department of Geography (Ramsli, 1974).

Avalanche legislation and regulation in Canada are a provincial and a local concern. In British Columbia the ministry of the attorney general, the ministry of highways and transportation, and the ministry of municipal affairs play key roles in legislating and regulating avalanche hazards. The ministry of the attorney general is involved in avalanche management through the administration of the Land Title Act, RSBC 1979, especially with regard to subdivision plans. If the land in question is in a municipality, subdivision approval is a municipal concern; if, however, the land is rural, subdivision approval is obtained through the ministry of highways and transportation. For rural land, subdivision plans must be accompanied by topographic details that include environmental impact assessments. Within the Land Title Act, subdivision plans can be refused if "the land is subject, or could reasonably be expected to be subject, to flooding, erosion, land slip or avalanche" (B.C. Government, RSBC 1979, C.219, Sec. 86). Specific reference is made to avalanches; thus, both municipalities and highways have the mandate to control development and refuse it where potential avalanche hazards exist.

The ministry of municipal affairs is also involved in avalanche management: under the Municipal Act, RSBC 1979, municipal councils have the power to relocate and close municipal highways, develop community plans, regulate sitings of buildings, and regulate land use through zoning. Municipal councils also have the power to restrict specific uses within a particular zone, and it is this potential management strategy that is particularly useful in preventing large-scale loss of life and property damage from avalanche hazards. Yet as of 1984, zoning had been ad hoc and infrequent (McFarlane, 1984).

In the United States the onus to zone lies mainly with state and local governments, rather than federal agencies (Niemczyk, 1984). Most progress has been made in Colorado, where in response to an increase in land speculation and recreational development in the Rocky Mountains (Ives and Krebs, 1978), House Bill 1041 was legislated in 1974. This bill was concerned with hazard zoning for land-use planning; its adoption by local governments was voluntary (Rold, 1979), but counties were required at minimum to map their geological hazards and to use the mapping as a basis for land development approvals. Before development controls could be implemented, it was first necessary to identify buildings already situated in avalanche paths; in Vail, for example, 40 such structures were identified (Ives and Krebs, 1978). After the initial identification procedure, a few developers responded by introducing design changes into partly built structures located in avalanche paths; others did not. However, since 1974, new proposals for development have been subjected to strict building codes and regulations.

Alta, Utah, also has an avalanche zoning plan. It is administered by the Salt Lake County Planning Commission, which controls development in avalanche zones through building permits (Tesche, 1977). Another example is Ketchum, Idaho, where about 35 residential lots with buildings were found to have avalanche paths directly affecting them. Restrictions were placed on subsequent buildings through special design specifications; defense structure requirements; confining residence to the period April 15 to November 15; property subdivision prohibitions; and, in addition, a requirement to notify tenants, lessees, real estate agents, and sellers of any avalanche hazards (Mears, 1980). Similarly, in Placer and Nevada counties in the Sierra Nevada of California, building ordinances enacted in 1982 for avalanche zones place engineering requirements on new construction, reconstruction, and expansion of existing structures and require written notification to renters and buyers (Penniman, 1989a).

There are no land-use restrictions for buildings in avalanche paths in Alaska (Tesche, 1977; Hackett and Santeford, 1980; Mears, 1980), although omnibus legislation (Senate Bill 301) passed in May 1980 relates to avalanche warning and control systems (James, 1981). Through Bill 301, Alaska's Department of Public Safety is mandated to forecast and control avalanche hazards and coordinate an avalanche information program and "to assist local governments and state agencies in identifying hazardous avalanche zones and in developing snow avalanche zoning regulations" (James, 1981). Thus, a vehicle has been established for future avalanche zoning policies in Alaska. Unfortunately, it is little used.

HAZARD DELINEATION

Avalanche legislation usually prepares the way for avalanche zoning, which in turn subdivides the land so as to enforce building restrictions. Criteria vary from country to country.

Three types of hazard maps are distinguished in Norway: hazard registration maps, geomorphic hazard maps, and hazard zoning maps (Hestnes and Lied, 1980). Hazard registration maps detail historically known avalanches from literature, documents, interviews, and field work. Geomorphic hazard maps add the results of geomorphic investigations, while hazard zoning maps show both actual and potential paths, together with mathematically or statistically derived runout distances. Typical survey maps give general information about hazards at a scale of 1:50,000. Detailed maps at 1:5,000 scale have high accuracy but demand comprehensive and time-consuming field work. Hazard zoning maps include an estimate of potential risk (i.e., future natural hazard activity and damage). Acceptable levels of risk for housing are evaluated in comparison with other types of social risk. The proposed highest tolerable risk level for damage to dwelling houses in Norway is 3×10^{-3} per year (Hestnes and Lied, 1980), well above the proposed international standard for "planned activities" of 1×10^{-6} (Starr, 1969).

Mapping for communities, highways, construction sites, and power lines is carried out by the Norwegian Geotechnical Institute (H. Norem, Norwegian Geotechnical Institute, Oslo, written communication, 1986). Over 1,000 paths had been recorded in detail by 1980, but the objective is to cover 100,000 km², nearly a third of the nation's land surface. Some recent mapping has been done based on topographic information alone, using computer-based digital terrain models to identify avalanche starting zones and to establish runout boundaries with empirically based modeling laws (Toppe, 1987).

In the U.S.S.R. a national inventory has been completed. In this inventory, snow avalanche hazards have been placed into the second most significant group of hazards, classed as destructive natural phenomena that seldom cause loss of life but that result in significant damage to the economy, especially industry (Gerasimov and Zvonkova, 1974). Because avalanches occur in approximately 20 percent of the land area of the U.S.S.R. (Tushinsky et al., 1966; Akifyeva et al., 1978), zoning is important. Russian criteria for zoning are based on vegetation characteristics. It is recommended that no construction be allowed in avalanche "natural-territory complexes of meadow and subalpine elfin formations," whereas construction of summer buildings should be permitted in avalanche "natural-territory complexes with mixed forests and pine forests of various ages in which recurrence intervals range from fifty to three hundred years" (Akifyeva et al., 1978). This type of zoning explicitly takes into account the frequency of an avalanche, its size, areal extent, impact pressures, and seasonality.

Avalanche hazard mapping at a scale of 1:50,000 was initiated in France in 1964 under the direction of the minister of water and forests (Cazabat, 1972). The scale of mapping was changed to 1:20,000 in 1970, and Cartes de Localisation Probable des Avalanches (CLPA) maps were produced for an area of 7,000 km² in the Alps and Pyrenees, using photointerpretation, field investigations, and assemblages of local witnesses confronting one another—thus testing the accuracy of recollections against each other or against historical documentation (de Crecy, 1980; Brugnot, 1987). These CLPA maps, prepared in a novel fashion, were distributed to civic authorities responsible for public safety and land-use planning.

In addition, French law obliges the government to define the limits of zones subject to natural risk before a building permit can be obtained. Consequently, in 1974, Plans des Zones Exposées aux Avalanches (PZEA) maps at a scale of 1:2,000 or 1:5,000 were introduced as a legal document to cover areas where town planning is envisioned or is in progress (de Crecy, 1980). Three zones are delimited by specialists in PZEA zone maps. The red zone indicates a high degree of danger and no construction is permitted. The blue zone is an intermediate hazard area that includes extremely rare and yet severe avalanches with return periods greater than 300 years, as well as less severe avalanches with return periods every 30 to 50 years. Construction in the blue zone is permitted only under conditions established by specialists, which include evacuation when required, reinforcement specifications for buildings, protective structures (forests or man-made defenses), and/or annual inspections. A building in the blue zone may be denied windows on walls facing uphill, and the pitch of the roof may be specified (de Crecy, 1980). Research on mathematical avalanche models was undertaken to provide tools to assist in the delineation of blue zone boundaries and to provide criteria for structural design. The third zone, white, is an area that is most likely safe from avalanches and where construction thus is not restricted.

The PER risk map, a legal document defined by 1982 and 1984 laws, follows some of the guidelines of the existing PZEA procedure (Brugnot, 1987). Red, blue, and white zones are again established; these have implications concerning insurance, as discussed below in the section on Insurance and Disaster Relief.

Zoning decisions are nevertheless complicated, involving local governments and appro-bation of the communities involved. Although a general technical handbook describing the state of the art in natural hazard protection was prepared for distribution to local author-ities (Délégation aux Risques Majeurs, 1985), additional technical expertise is recognized as necessary. A population of experts is likely to cause discrepancies and inconsistencies; this is considered a problem, since the PER maps are government endorsed and the aim is to treat all citizens in an identical fashion with regard to building restrictions that reflect natural hazards (Brugnot, 1987). In early discussions concerning the PER procedure, some government officials expected such problems to disappear through advances in technology. This opinion seems unrealistic, despite the recognized need for continued research and de-velopment of scientific tools, such as dynamic modeling, an avalanche data base, and expert systems (artificial intelligence). Work on the latter topic commenced in 1985 (Brugnot, 1987; LaFeuille et al., 1987).

Switzerland uses two types of hazard maps: avalanche zone plans and avalanche hazard maps. Avalanche zone plans are legitimized by avalanche zoning laws and are legally binding; moreover, they are a component of the building code (Frutiger, 1980). Avalanche hazard maps, on the other hand, are not legally binding and are merely tools to assist the decision maker in regulating land use. The distinction between the two types of maps, made in 1975 by the Federal Bureau of Forestry (Frutiger, 1980), is used in the following discussion.

Under the Swiss color zoning scheme, impact pressure and avalanche frequency are used to quantify the degree of risk. This is not a purely scientific question but also a political and psychological one; authorities have to decide the level of risk that should be accepted (Buser et al., 1985). The maps are compiled at scales of 1:25,000 and 1:10,000 (Kienholz, 1978; Frutiger, 1970, 1972; Buser et al., 1985; cf. Mears, 1979, 1980). Red zones constitute the areas of highest risk. Avalanches are either powerful (impact pressures greater than 30 kN/m², or approximately 0.5 psi) with a return period of 300 years or less or frequent (return periods up to 30 years), irrespective of intensity (Buser et al., 1985; Bundesant für Forstwesen, 1984). New buildings and winter parking lots are generally not allowed in this zone. Blue zones have dynamic pressures less than 30 kN/m² with return periods of 30 to 300 years. Residential development is permitted if it is protected by avalanche defenses or if construction meets design specifications to resist avalanche forces. The specifications can include such criteria as building strength, materials, shape, size, spacing, or function (Kienholz, 1978). Churches, schools, hospitals, lodges, and other public places are not permitted in blue zones. The white zone is beyond the limit of design avalanches, though not necessarily outside the range of all possible avalanches.

An optional yellow zone has been used to define an area where avalanches are rare or where air blasts occur. Buildings in this zone must conform to building code standards. This area can be impacted by powder avalanches with dynamic pressures of 3 kN/m² (approximately 0.05 psi) or less and a return period greater than 30 years and/or by rare flowing avalanches with return periods exceeding 300 years. The latter are not well understood, and the criteria for determining their occurrence are subject to question (Frutiger, 1980; Brugnot, 1987). No building restrictions related to avalanche hazard are prescribed. Swiss research continues on natural avalanche dynamics and on avalanche modeling, in order to provide better tools to meet the requirements of hazard zoning (Gubler, 1987, 1989).

In Canada the avalanche hazard line is considered the boundary of large infrequent avalanches, and development is defined as being inside or outside this line. The line, which corresponds to the boundary of the blue and white zones of the Swiss system, indicates how far extreme avalanches can reasonably be expected to travel. Its establishment does not account for return intervals and impact pressures. Occasionally, within active sites, additional categories are used to include frequent flowing avalanches with return intervals of less than 30 years, infrequent flowing avalanches with return intervals exceeding 30 years, and infrequent powder snow avalanches and wind blasts. In addition, a safety distance of 50–150 m (150–450 ft) is added onto the known boundary of the runout zone (Freer and Schaerer, 1980). These hazard lines are recommendations only, and their implementation is at the discretion of the approval officer. If the developer does not agree with the zoning, an appeal may be made to the approval officer or the courts. Runout distances are established by terrain and vegetation analysis and by mathematical modeling. Base maps in Canada are at 1:50,000, a scale inappropriate for detailed avalanche studies. Aerial photographs are used extensively.

No uniform policy exists in the United States. In Colorado, House Bill 1041 (1974) requires local communities to compile hazard maps. Since some of the work subsequent to this legislation was delegated to the Colorado Geological Survey while other work was given to private consultants (Rold, 1979; Ives and Plam, 1980; Mears, 1980), there are a number of different types of hazard maps and zoning plans, all with different criteria. Avalanche hazard maps compiled by the Colorado Geological Survey are at a scale of 1:24,000 and are published at a scale of 1:50,000 (Mears, 1979); they define high hazard, moderate hazard, and no hazard areas based on impact pressures and avalanche return periods. "No hazard" is assigned to areas considered free of avalanches or with return periods up to one or two centuries, or where air blasts might occur, and is ignored for planning purposes. Recommendations are made that no permanent residences be allowed in high hazard areas and that engineering design precautions be adhered to in moderate hazard areas (Mears, 1979).

The Institute for Arctic and Alpine Research has also been involved in avalanche hazard mapping in the United States (Ives and Bovis, 1978; Ives and Krebs, 1978; Ives and Plam, 1980). In the institute's unpublished 1:24,000 maps, avalanche paths are designated according to whether they are active or potential. While these maps give some idea of the location of hazard, they do not have the technical standards required for assisting in regulation of specific development in mountain areas. Detailed mapping with impact pressures and return probabilities is desirable for land-use planning in mountain environments.

In Alaska, maps indicating high hazard, potential hazard, and no hazard have been compiled at scales of 1:250,000 and 1:53,000 (Hackett and Santeford, 1980). These scales are inadequate for detailed planning but represent a step in the right direction. Detailed maps are available for some areas.

With respect to avalanche zone plans, similarities are shared by Vail, Colorado, and Ketchum, Idaho. Both communities have a color scheme to differentiate degrees of avalanche hazard, white being nonhazardous. Red zones correspond to high hazard areas in which "impact pressures on a flat surface normal to the flow" exceed 600 lb/ft^2 and/or avalanche return periods of less than 25 years (Mears, 1980). Residential construction is not permitted in red zones. In blue zones, impact pressures are less than 600 lb/ft^2 and avalanche return periods are approximately 25 to 100 years. Building in this zone is permitted if design by a registered engineer provides for avalanche forces. This modification to the Swiss definitions recognizes the uncertainty in specifying long return periods in a region of short observational records. Responsibility for providing safe design lies with the property owner or the owner's consultant, not with the community. Another plan, developed for Ophir, Colorado, uses slightly different frequencies and impact criteria. The important factor in all these cases is the incorporation of avalanche zone plans into the city's land-use ordinances.

Zone plans for Placer and Nevada counties, California, illustrate some zoning problems for two geographically similar areas faced with identical avalanche hazards. The zone plans are based on a single hazard study conducted for both counties at the same time, by the same consultant, using identical procedures (Penniman, 1989a). Hazards were defined by a red zone (high hazard, occurrence probability for a damaging avalanche 1:20); a blue zone (moderate hazard, probability 1:20 to 1:100); a yellow zone (low hazard, probability less than 1:100); and a white zone (no hazard). The Placer County ordinance places engineering requirements on new construction in red and blue zones and notification of hazard to property users in yellow zones. In contrast, Nevada County requires engineering measures in red zones and only written notification in blue zones and treats equally the yellow and white (no hazard) zones. The wisdom of ignoring yellow zones can perhaps be challenged on legal and ethical grounds, and subtle issues arise when property owners appeal hazard ratings based on a site-specific study. As in all cases involving municipalities with avalanche problems, policymakers are in the unenviable position of deciding whether the public good is best served by warning the public of a potential danger or by trying to maintain property values until the danger is proven.

INSURANCE AND DISASTER RELIEF

Avalanche insurance is not available in many countries; disaster relief is often the alternative. A few exceptions are worth noting.

Norway is one of the few countries in which private insurance against avalanche damage can be obtained. However, private insurance is generally not necessary because avalanche damage insurance is compensated by the Norwegian National Fund for Natural Disaster Assistance (Ramsli, 1974; Hestnes and Lied, 1980). This organization has been active in preventing development in high-risk areas, and has supported pilot hazard mapping projects by the Norwegian Geotechnical Institute. From 1962 to 1971, Norway was subjected to an estimated $1.6 million worth of private property damage from avalanches (Ramsli, 1974).

Although insurance was not available earlier in Switzerland (Perla and Martinelli, 1976), in recent years it has become an increasingly popular option. Proposals are now being tendered for using insurance companies to enforce building regulations, and it has been suggested that buildings constructed in hazardous places be denied avalanche insurance (Frutiger, 1980). Such a strategy might force cantons and communes to implement land-use zoning and thus ensure development only in avalanche-safe areas. In some cases insurance companies are a cantonal institution, and their guidelines carry official legal authority.

The most elaborate and well-defined natural hazard insurance system is in France; it is regulated by the PER risk map concept (Brugnot, 1987). Accordingly, for future or existing construction in red zones, insurance companies may refuse protection; in the former case the possibility also exists for legal prosecution against the community. For new construction in a blue zone, no insurance company can refuse to insure the property if the provisions of the PER have been followed regarding reinforcement or protective structures. For preexisting buildings in a blue zone, the insurance company cannot refuse protection, but the owner has a 5-year period in which to comply with PER requirements.

Several aspects of the French system are particularly important: first, natural hazard insurance is implicit, which signifies that no additional insurance cost applies to blue zone properties. Thus, every French household contributes to natural hazard indemnification. The national increase in insurance fees due to natural hazards was about 8 percent in 1985 (Brugnot, 1987). Second, any refund of damages is conditioned by acknowledgment of the existence of a natural disaster situation by a government panel. To date, practically every avalanche damage case presented for indemnification has been accepted, even disputed ones. Because the most costly part of the insurance scheme, by far, concerns flooding, other hazards have been considered leniently. Most avalanche cases admitted as disaster situations were defined in terms of large snowfalls, but exceptional snow metamorphism also was cited as a criterion to satisfy the specification of a disaster situation.

Avalanche insurance in New Zealand can be obtained through a national natural hazard insurance policy (Olshansky and Rogers, 1987). Any building covered by fire insurance is automatically charged an additional $0.05 per $100 of coverage; the funds are split—90 percent for the Earthquake and War Damage Fund and 10 percent for the Extraordinary

Disaster Fund, both administered by the Earthquake and War Damage Commission. To qualify for natural hazard insurance, a property owner must have safeguarded the property against a "normal" or "reasonably expected" event. Thus, unless building codes have been met, coverage may be refused or may be assigned a more costly premium. This insurance, however, is only for the "abnormal" event; the "normal" event must be covered through private insurance even though private insurers currently do not like to insure high-risk properties (O'Riordan, 1974).

In the U.S.S.R., insurance against avalanche damages can be obtained from the "Gosstrakh" (Gerasimov and Zvonkova, 1974). This government-administered insurance agency has a mandate to cover losses from natural hazards. The Soviet government compensates owners for all damages to buildings and animals caused by "natural processes not peculiar to a given region" (i.e., abnormal occurrences—similar to New Zealand's restrictions). Thus, areas that can be insured against avalanche damage are, for example, forest areas below avalanche runout zones never before impacted by avalanches.

In Canada the costs of disaster relief are shared by the federal government and the provinces, since private avalanche insurance is not available. The national agency that administers disaster relief is Emergency Planning Canada (EPC); its involvement in avalanche disaster has so far been limited to large-scale events. In the absence of relief funding from EPC, other federal and provincial administrations such as the British Columbia Provincial Emergency Program have absorbed the costs.

Private insurance against avalanches is generally not available for property owners in the United States. Nevertheless, under lenient interpretation, repairs due to an airborne dry snow avalanche in March 1962 in Juneau, Alaska, were covered by homeowners' insurance policies; the insurance company adjustors determined that the damages were caused by the wind (Hart, 1972). A national landslide insurance fund has been proposed but not enacted, accompanied by a program for mapping hazard zones and determining actuarial rates (Olshansky and Rogers, 1987). National hazard relief funds are administered by the Federal Emergency Management Agency (FEMA), and funding of emergencies is shared by the federal government and the states in question. However, no avalanche damage has yet been covered by FEMA funds.

FEMA has struggled with the dilemma of how to spend disaster funds not only to aid victims but also to encourage mitigation efforts. With respect to landslides, FEMA has wavered between a liberal policy of paying for stabilization and reconstruction of public infrastructures, which does little to discourage development in hazardous areas, and a strict policy of allowing only emergency repairs (Olshansky and Rogers, 1987).

Finally, personal insurance for skiers (covering, for example, search and rescue costs and legal fees related to injuries to other skiers) is available worldwide from Carte Neige, through La Fédération Française de Ski (Wells, 1987).

COMMENTS

1. Legislation regulating land use in avalanche hazard areas is most restrictive in Switzerland, France, and a few regions of the United States (particularly Colorado). There is a lack of overall legislation requiring local communities to zone for avalanche hazards in Canada, Norway, and most locations in the United States, although in Norway the National Fund for Natural Disaster Assistance has been active in preventing development in areas at

risk. Legislation is desirable that requires properties with known avalanche hazards to be registered, since such legislation offers protection to a buyer.

2. Avalanche hazard maps and avalanche zone plans require the level of clarification specified in Switzerland or France; terminology and definitions, often not used uniformly, can lead to ambiguities. Additional clarification of terms would help differentiate legislated versus nonlegislated avalanche hazard mapping.

3. The maps used for avalanche hazard mapping are often too small to clearly specify hazard boundaries. Maps at a scale of 1:50,000 do not provide enough detail to be useful in land-use planning, where detail is critical. Thus, 1:25,000 is the minimum scale acceptable for general avalanche zone plans, while 1:10,000 to 1:2,000 is desirable for greater accuracy.

4. European nations with modern zoning plans invariably also carry out effective programs of research in order to develop, calibrate, and modernize tools necessary for the definition of zone boundaries and the establishment of design criteria. No such research is currently supported in the United States.

5. A number of countries with avalanche problems have insurance coverage, with the French system in many respects the most novel. Insurance policies can be effectively used to support building regulations if individuals must meet required zoning and design specifications to qualify for insurance.

6. Insurance systems demanding extreme or abnormal events, as in New Zealand and the U.S.S.R., serve a restricted if occasionally important role. The French definition of disaster conditions provides readily available protection to the public yet also provides the incentive to build in safe areas and to use adequate protection.

7. Comparison of the United States with other countries reveals that overall avalanche management has, at least in the past, been accorded some regional emphasis. Options that have not been adequately explored include avalanche insurance and land registry of avalanche-threatened property. It should also be noted that the current use of zoning strategy is subject to criticism, for it is carried out on an ad hoc basis rather than comprehensively. These are aspects of avalanche management that would benefit from an overall strategy. Additional aspects are discussed in the next chapter.

5
Avalanche Control

Control techniques used in the United States are comparable to those used in other industrialized mountain countries. However, there is a growing disparity between the type and extent of techniques used in the United States and those used in such countries as Switzerland, France, and Austria where a long-term commitment to the reduction of avalanche hazards has achieved greater progress in avalanche control and a higher priority for public safety.

The objective of avalanche control is to reduce or eliminate the hazard from potentially destructive avalanches. Methods for accomplishing this include (1) active methods, which involve systematic attempts to artificially trigger small nondestructive avalanches as a means of reducing the hazard as well as to test the accuracy of avalanche hazard forecasts, and (2) passive methods, which include anchoring or modifying the snow in avalanche starting zones so as to eliminate the release of large destructive avalanches and the construction of various structures to divert or dissipate the force of an avalanche in track or runout zones.

ARTIFICIAL RELEASE OF AVALANCHES

Avalanches may be initiated by detonating high explosives either in or above the snowpack. When such artificial triggers produce avalanches, impressions about snow stability can be ascertained, and options for avoiding the consequent hazards can be formulated. When efforts to trigger avalanches fail, however, it should not be concluded that the snowpack is necessarily stable (Gubler, 1983; Pratt, 1984; Penniman, 1987).

Mechanical shear loading to the snowpack in starting zones can be accomplished with or without explosives. Explosives can be used to drop cornices or release smaller sluffs from above onto large avalanche starting zones (McCarty et al., 1986). This safe and effective way of applying large shear loads to a slope is often helpful in determining the stability of the snowpack and in triggering avalanches. Under certain conditions, cornices can be safely kicked loose by experienced technicians to test the stability of lower slopes. Alternative experimental methods for releasing avalanches include gas detonated above buried canisters (e.g., GAZ.EX), air-bag inflation, and "seismic exploration" air guns (LaChapelle,

1977, 1978; Penniman, 1989b; Tremper, 1990; D. Abromeit, U.S. Forest Service, written communication, 1990).

Since 1933 the most versatile and practical techniques for artificially triggering avalanches utilize various forms of high explosives to induce a shock wave into the snowpack (Fraser, 1966; Seligman, 1962). Over 100,000 explosive charges are detonated annually for avalanche control (Perla, 1978b). The equivalent of 1 kg (2.2 lb) of TNT has been established by tradition as the standard charge for testing snow stability, but larger charges can and often are used when necessary, and smaller charges may be adequate for thin new snow (Perla, 1978b). The best results with explosives are achieved from detonations that occur 1–2 m (3–6 ft) above the snow surface or on rock surfaces near the target areas in starting zones. Correct placement and correct timing of explosive detonations are critical to their effectiveness (Gubler, 1977, 1983; LaChapelle, 1978) and are often a matter of local experience.

Techniques that utilize explosives have been reasonably safe and effective for the majority of snow conditions when strict safety precautions are observed and generally accepted control procedures are followed. However, with certain conditions, such as wet snow, explosives have often been unreliable. Some serious safety problems remain unresolved, as will be noted later; liability issues are discussed by Fagan and Cortum (1986).

A variety of delivery systems are currently in use, the most common of which is hand delivery. This technique, widely used at U.S. ski areas, requires avalanche control technicians to ski or walk to predesignated delivery sites and physically throw charges into known avalanche starting zones. Costs are comparatively low when a large number of avalanche paths are concentrated in easily accessible areas, and the placement of explosives can be widely adjusted to achieve greater effect in various snow deposition patterns. Disadvantages of hand charging are that the procedure cannot be readily performed at night or during extreme storms and the avalanche control technician may be exposed to hazardous conditions. Suspending the charge at the desired height above the snow surface or on rock surfaces is also impractical without significantly increasing the cost and time necessary to conduct operations. Experiments in Switzerland with booms that swing out over a starting zone to suspend a charge have had some success; apart from the Alpental ski area in Washington, none are in current use in the United States.

The hand charge is currently the predominant explosive system for avalanche control in terms of the number of explosive charges. The hand-charge system, ignited with a pull wire, seems to be relatively safe, as few explosive accidents have occurred despite wide variation in the types of explosives used and the broad range of deployment conditions. Areas under U.S. Forest Service (USFS) permit were, at one point, required to develop safety plans for training personnel in the use of hand charges, but there has been reluctance among some suppliers of explosives to be involved with hand-thrown applications due to a lack of studies about the reliability of the assembled hand-charge configuration as well as distrust toward departures from standard procedures used in normal blasting practices.

Two hand-charge accidents in 1973 at Mammoth Mountain, California, probably involved some aspect of the pull-wire fuse igniter attachment and led to formal testing of the hand-charge system by the Naval Weapons Center at China Lake, California, at the request of the USFS. Testing revealed that the system in use at the time could experience detonation from electrostatic fields, thus indicating the need for a grounded or nonconductive fuse. The primary cause of the accidents was apparently poor operational procedures. The test report concluded that the USFS should institute a procedure for certifying proficiency in handling of specific detailed safety instructions: "These must be specific, not broad, platitudes such as 'the operation shall be conducted in a safe manner' " (Austin et al., 1974). It is difficult to prepare specific safety guidelines without referring to a specific hand-charge system (Perla, 1978b).

Cable delivery systems are little used in the United States (Dombroski, 1988). These systems are being installed throughout much of Europe (Gubler, 1983) and in some parts of Canada. Of Austrian and German origin, over 120 cable explosive transport systems are now used in France alone for ski areas and transportation routes (Brugnot, 1987, 1989; Borrel, 1987; Rapin, 1989).

Using manual or powered drives, cable delivery systems transport explosive charges to avalanche starting zones on a cable tramway. Once in position, sophisticated remote-control carriers automatically lower charges to the appropriate height above the snow surface and then detonate them. Cable systems more than 6 km long sometimes require computer-aided motor drive and radio-signaled explosive control (Brugnot, 1987). These systems can deploy several carriers at once, thereby saving time, and can be operated at night and in poor visibility from a safe location with maximum effectiveness, allowing inaccessible or dangerous starting zones to be remotely accessed. Depending on their design sophistication, cable systems can appear relatively expensive to build and to operate, yet they seem to be cost-effective. To some, there are aesthetic problems—the systems do not beautify the landscape. Regulations in France require the retrieval of explosive charges after a 30-minute delay if firing has failed (Brugnot, 1987). This creates operational difficulties but is in the interest of public safety.

Helicopters can be used to deploy explosive charges by aerial bombing. They are also used to transport control technicians to otherwise inaccessible terrain for hand-charging operations. In the United States, Federal Aviation Administration regulations govern the operation of helicopters for the transportation of explosives and for aerial bombing operations. In the United States and Canada, helicopter delivery is commonly practiced by helicopter ski companies and by mining and construction companies for short-term projects (Gmoser, 1978; Perla and Everts, 1983). The method allows a very accurate and fast inspection of starting zones and placement of charges. Helicopter flights are, however, limited to favorable weather conditions, and explosive charges cannot be suspended above the snow surface or placed on rock surfaces to achieve maximum effect. In fact, because helicopter-dropped charges penetrate deeply into the snow, heavier than normal charges must often be used to gain the same effect as with a standard hand-thrown charge. Although the hazards of hand charging do not exist with aerial bombing, flying in mountainous terrain can be equally dangerous.

Preplanted explosives systems have not been used much in the United States and have seen only limited use elsewhere. These systems have the advantage of being installed during good summer weather, and because they are detonated remotely, there is virtually no hazard to technicians. While the systems can provide control for otherwise inaccessible starting zones, they are very susceptible to mechanical failure due to stress on components buried by snow. Another disadvantage is that only a limited number of charges can be placed and only in fixed positions. The relative cost of installing remote systems is high, and the effectiveness of detonation in deep snow is questionable (Perla and Everts, 1983).

In the United States the use of artillery is a predominant method of avalanche control. The advantage of artillery is that it can be fired at any time of the day or night, regardless of weather. Artillery rounds can also be fired into rock surfaces near target starting zones of weather. Artillery rounds can also be fired into rock surfaces near target starting zones for better effect. As with aerial bombing, rounds that must be shot into the snow usually detonate below the surface and can be less effective in deeper snowpacks (Perla, 1978b). The resulting shrapnel can be a hazard, and overshooting is always a possibility, with the accompanying threat of property damage and injury.

Both military-produced artillery and civilian-produced artillery are widely used by U.S. ski areas, highway departments, and industry. Military artillery pieces include 75-mm and 105-mm recoilless rifles (RR), the 75-mm mountain howitzer, and the 105-mm howitzer. Field tests of 106-mm recoilless rifles are scheduled for the 1989–1990 season (Penniman, 1989b; D. Abromeit, U.S. Forest Service, personal communication, 1990). The explosive content of ammunition for these weapons varies from about 0.7 to 3.5 kg (1.5 to 8.0 lb) of high-speed explosive.

The supply of ammunition for the World War I 75-mm howitzers is limited, although in the past, at critical intervals, ammunition supplies have been "discovered." Despite its age, this weapon remains one of the more popular rifles in use, especially in places where high accuracy and reliability are essential because of proximity to populated areas.

In 1950 recoilless rifles were made available by the U.S. Army, which helped reduce dependence on the dwindling supplies of World War I ammunition and replacement parts for the 75-mm howitzers (LaChapelle, 1956, 1962). The recoilless rifles are lighter than the howitzers, and because of their low recoil they allow lighter support structures and permanent gun emplacements. Permanent gun emplacements in turn permit instrumental alignment for blind firing during periods of poor visibility.

The recoilless rifle is the principal type of artillery currently used for avalanche control. Some areas are using this weapon by choice because its shorter range reduces the chance of overshoot into populated areas. Recoilless rifles require frequent vent inspection and vent replacement. A shortage of adequate spare vents is considered to be a major problem for some users. Once again, however, the major problem facing users is the possibility that aging ammunition may be withdrawn from the program, as it was in the summer of 1985 (Abromeit, 1988; D. Bowles, Utah Department of Transportation, personal communication, 1986). Inspections by the U.S. Army of ammunition prior to shipment are made on a lot-sample basis to ensure that the ammunition meets acceptable standards of use. To avoid interruptions in critical avalanche control programs, users have tried to maintain large ammunition inventories. While this provides a longer-term supply, it fails to address the level of inventory control and inspection formerly guaranteed by Army military storage and testing procedures.

Current (1989) estimates indicate that for users of the 105-mm RR there is at least an 8-year supply of serviceable rounds and an additional 6 to 7 years of "unusable but repairable" rounds (Penniman, 1989b). There is only about a 4-year supply for the 75-mm RR. For these reasons the U.S. Army has now relaxed its prohibition on civilian use of the 106-mm RR. However, this weapon is also out of production, and its use represents only a temporary solution.

The only civilian artillery piece being used in this country, the "Avalauncher," is produced in California by R. C. Peters Avalanche Control Systems. This device is a compressed air cannon that propels a 1-kg, rocket-shaped projectile a distance of 1 km (Atwater, 1968). The projectile detonates on impact and throws no shrapnel. Its range and accuracy are inferior to those of military weapons, but use of the Avalauncher could increase because users have been warned that military ordnance will be depleted within a few years at current than do conventional military weapons. However, there are safety concerns, and production problems plague the manufacturer, leaving Avalaunchers and projectile parts in chronic short supply. Other substitutes for military weaponry have been proposed, but none have been developed (Perla, 1978b; Penniman, 1989b).

CONTROLLING THE USE OF EXPLOSIVES

Prior to World War II the USFS pioneered the use of explosives for avalanche control. Subsequent efforts by the USFS to obtain military weapons for avalanche control came shortly after World War II, when the first artillery tests for avalanche control were conducted with French 75-mm howitzers at Berthoud Pass and soon after at Alta (Kalatowski, 1988).

An immediate result of these successful tests was the development and acceptance of a guideline "Memorandum of Understanding" between the U.S. Department of Agriculture and the U.S. Department of Army (J. Herbert, U.S. Forest Service, personal communication, 1986; Kalatowski, 1988). This memorandum defined the roles and responsibilities of the two departments in what would become the weapons program. Under the terms of agreement, the U.S. Army would supply surplus weapons and ammunition, with repair and training support, to local USFS offices. The USFS would administer the program and assume responsibility for training gun crews, operating the program, and maintaining public safety. The memorandum had the effect of making the USFS and the U.S. Army partners in selected areas and created a protective "umbrella" to spread the risks.[1]

During the early period, avalanche control artillery was fired by National Guard gun crews, with target selection by USFS snow rangers. While the firing by National Guard crews was highly professional, the lead time for weapons deployment was immoderately long compared to forecast lead time. By 1966 the roles were more clearly defined. In areas having a high hazard, defined as Class A, the USFS would provide snow rangers with avalanche forecasting and artillery control expertise. Areas with less serious avalanche problems were classified as Class B or Class C. Class B areas were monitored and assigned direct USFS control if they failed to provide adequate avalanche protection for the public. Class C areas did not require rifles or direct snow ranger supervision. Over time a gradual shift was made to the employment of ski-area personnel as gun crews, and the USFS's role was reduced to administrative monitoring with little hands-on gun time (D. Bowles, Utah Department of Transportation, personal communication, 1986).

In some instances, weapons control programs have been developed by state agencies. Highway departments in Alaska, California, Colorado, and Washington have developed successful avalanche control programs similar to those of USFS-administered ski areas (LaChapelle, 1962). The state governments entered into local agreements, usually involving both their National Guard and the U.S. Army, to supply weapons and support.[2] Generally, the resulting programs have faced problems similar to those encountered in USFS programs, mainly in the areas of spare parts, ammunitions availability, dud disposal, and gun crew training.

MAJOR PROBLEMS IN THE USE OF EXPLOSIVES

Unexploded charges (duds) represent one of the most serious operational problems facing explosives control programs, particularly for the artillery program (Abromeit, 1988). Self-destruct capabilities are not normally built into military warheads.

In many areas dud rates of 2 to 5 percent are common (D. Bowles, Utah Department of Transportation, personal communication, 1986). In most instances the rounds are fully armed but fail to explode on contact with the snow. Armor-piercing rounds are less sensitive and generate a larger percentage of duds. Rounds of high explosive plastic tracers (HEPT) have shown a dud rate up to 30 percent [D. Abromeit, cited by Penniman (1989b)]. Also, recoilless rifles yield a substantially increased dud percentage when used at over half the maximum range (Perla, 1978b). This increase is due to the influence of trajectory; with flat-trajectory grazing shots into soft powder, some projectiles skip back into the air and continue their flight to some other landing site. Unexploded rounds have been found at the maximum range for the weapon, which for the 105-mm recoilless rifle and the 75-mm howitzer is over 8 km (5 miles). Most duds fall into remote and inaccessible areas, but despite a low encounter probability many are found each year (Perla, 1978b).

Military ordnance experience suggests that 10 percent of duds detonate spontaneously (D. Bowles, Utah Department of Transportation, personal communication, 1986). The remaining 90 percent remain fully armed in some unknown state of sensitivity. Because military ammunition is well constructed and sealed to withstand long-term exposure to extreme environmental conditions, duds may remain operational for years. Most areas using weapons have had this problem since the inception of weapons programs in the early 1950s, and the cumulative number of lost, fully armed, and sensitive explosive charges is probably in the thousands.

Immediate retrieval of unexploded charges is generally impossible, and therefore dud control is included in the spring cleanup operations for area gun programs. The recovery rate is no more than about 50 percent. If an average dud rate of 3 percent is assumed for an average annual national projectile expenditure of about 6,000 rounds (cf. Perla, 1978b), a recovery rate of 50 percent implies 90 lost rounds per year (D. Bowles, Utah Department of Transportation, personal communication, 1986). Since artillery has been used for over 30 years, perhaps 3,000 unexploded rounds could exist in the U.S. backcountry, threatening recreationists. The current tendency for 105-mm RR users to switch to the more abundant HEPT rounds should exacerbate the dud problem (Penniman, 1989b; D. Abromeit, U.S. Forest Service, personal communication, 1990). Similar problems may arise with the 106-mm RR. With increased urbanization and use of backcountry areas, the probability of dud encounters is expected to increase.

A second major problem is related to the Avalauncher, the only civilian artillery in use in the United States. Designed to meet a specific avalanche control problem, the Avalauncher provides short shots with low fragmentation and has the further advantage that duds are rapidly reduced to an inoperative condition by the open case design. Its initial development was supported by the USFS (Atwater, 1968), and further refinements have been made by the manufacturer. The projectile has a finned plastic case that can be loaded with any type of explosive, from cast primer to dynamite. Arming is achieved by air flow, removing an arming disk and safety pin as the projectile exits the barrel, and a magnetically retained firing pin initiates base detonation on impact. Ranges up to 1,500 m and beyond are possible, although the longer distances require a stronger projectile case to prevent case collapse in the barrel. English and French versions (Avalancheurs) are capable of distances up to 4,000 m (12,000 ft) (Brugnot, 1987); neither is used in the United States.

The Avalauncher has been widely accepted for avalanche control, despite little official recognition by such branches of the government as the Federal Alcohol, Tobacco, and Firearms Agency. The USFS lost interest in its development, though its view has been one of benign neglect, neither approving nor disapproving its use. Avalaunchers are used today at many ski areas under USFS permit. The history of its use is further obscured by scant documentation by either the manufacturer or users (however, see Ream, 1990).

Users have long recognized the Avalauncher's inherent defects, both in operational safety and quality control of the weapon and its design. Many users have implemented special operating procedures to resolve some of these problems and make its use somewhat safe (e.g., Marler and Fink, 1986). Because the Avalauncher is not a fail-safe system, the mechanism will fire with component failure in the firing or drive mechanism. As a result, air leakage can cause the mechanism to fire.

A fatal accident in Chile involving an Avalauncher led to an analysis of the device by the USFS San Dimas Laboratory (Spray, 1983), which concluded that the fusing system then used did not conform to standard ordnance practice. Such flaws in design would not be tolerated in military systems, which are under tight administrative control, with crews thoroughly trained and obedient to specific operating documents. No such control or documentation is guaranteed for civilian operation, and this represents a serious problem that should be addressed.

MECHANICAL COMPACTION AND DISRUPTION

In the United States the stabilization of snow in avalanche starting zones through compaction is performed primarily by recreation facilities personnel such as at ski areas. The process densifies the snow, adding strength and reducing the tendency for future slope weakening through temperature gradient metamorphism. Compaction is accomplished by "boot packing," skiing, or machine methods. Boot packing is performed by a group of individuals walking down a known avalanche path in early season. Though usually requiring only a single pass down the slope, this method is labor intensive, and in the United States has been limited to small, easily accessible avalanche paths. Ski compaction can be accomplished cheaply by skiing patrollers and by the public. Effective in breaking up cohesive snow slabs, the method is widely used in the United States and throughout the world. Machine compaction utilizes the weight of over-snow vehicles to densify the snowpack. The effect is similar to skiing but can be accomplished faster and with more uniform results. Nevertheless, machine compaction is not widely used, chiefly because of the inaccessibility of many starting zones and the current high costs of vehicles and cable belay systems.

STRUCTURAL CONTROL OF AVALANCHES

Structural avalanche control includes the natural or artificial anchoring of the snowpack in starting zones, structure-influenced redistribution of the snowpack in starting zones, and the structural protection of lives and property located in known or suspected avalanche paths.

Destructive avalanches may be prevented by retention structures that anchor the snow in starting zones. Such structures are most commonly used where avalanches threaten permanent facilities, towns, or roads. Provided snow depths do not exceed design parameters, such structures have proved effective, although their reliability may be questionable when snow cover is deep and poorly cohesive (Brugnot, 1987). The most common retention structures in use include snow rakes, snow bridges, and nets (Thomman, 1986; Lazard, 1986).

Earthen terraces and rock-filled steel baskets (gabions) have been used in the past but are seldom used now. Steel or earthen retention structures are usually designed as permanent structures, while wooden retention structures (rakes and bridges) are temporary and are used in conjunction with reforestation (Fraser, 1966; Jaccard, 1986; Montagne et al., 1984). In the latter case the maturing trees are expected to take over the job of anchoring the snow, and the wooden structures are either left to disintegrate over time or are removed.

While retention structures and avalanche path reforestation programs are used quite extensively in Europe and elsewhere, few have been instituted in the United States. Where snowpacks more than 4 m (12 ft) deep are common, as in the mountains of the Pacific coastal states, retention structures would have to be of massive proportions and are not economically feasible. In the Intermountain and Rocky Mountain states, however, where snow is less deep, retention structures could be installed more economically, but they would still be expensive and might encounter resistance on aesthetic grounds. In Switzerland structural control is subsidized by federal funds provided that building sites are selected in regard to avalanche zoning plans (Frutiger, 1972).

Under certain conditions the size and frequency of avalanches can be reduced through structures that alter storm wind patterns and thereby alter the deposition patterns of snow in starting zones. Such devices are usually used in conjunction with supporting structures and are not intended to eliminate the threat of avalanches, but rather to influence the amount and pattern of snow that accumulates in the starting zone. They are currently being used in a few parts of the United States.

One redistribution structure, called a "jet roof," acts as a "venturi" at the ridge line above avalanche starting zones (Perla and Martinelli, 1976). It reduces cornice buildup and causes wind-borne snow to deposit farther down the lee slope where inclinations are more gentle. Installation and maintenance may be expensive. Other redistribution structures include snow fences, which are usually located on flat ridge crests above starting zones or on windward ridges (Norem, 1978). Fences trap blowing snow in fetch areas before the snow can reach the starting zones or cornices above the starting zone. Redistribution structures are relatively inexpensive to build but have limited application. Their major disadvantage is that they are less effective when winds deviate from their prevailing directions or are absent altogether.

Retarding or catchment structures, such as mounds, ditches, terraces, and dams can be designed to foreshorten runout distances of avalanches. Mounds and terraces usually are used to stop, divert, confine, or slow moving avalanche debris in the lower track or the runout zone of avalanche paths; some have been used in Colorado and Alaska (LaChapelle, 1962; Mears, 1981; Yanlong et al., 1980). Dams are usually designed to stop debris and are normally located in runout zones. Numerous mounds and terraces may also be positioned above the dams to decrease the impact force on the main structure.

Retarding structures may be permanent, of earth, rock, and concrete construction, or may be large temporary berms of snow. An advantage of permanent retarding structures is their capacity to withstand tremendous impact forces. They require little maintenance once in place. A disadvantage is the short-term expense and the major visual transformation imposed on the landscape. Temporary structures made of snow are inexpensive to build and they disappear each summer, but they are not as strong as permanent structures, and maintenance is required after impact with major avalanches. Few permanent retarding structures have been built in the United States, but in Japan, Europe, and even in parts of South America they have been built with favorable results (Fraser, 1966; Mears, 1981;

Jaccard, 1986). Temporary structures built of snow have been successfully employed in California to reduce avalanche runout.

Other structures can be designed to protect permanent facilities, such as sheds, galleries, and tunnels to protect railroads and highways; berms of earth, concrete, or snow to deflect avalanche debris; and wedge-shaped walls that divert moving debris around specific structures or facilities (Fraser, 1966; Mears, 1981). In the United States, railroad galleries and tunnels have had success in reducing the number of avalanche incidents involving trains, but few structures have been constructed to protect highways from avalanches (LaChapelle, 1962; Mears, 1986). Dependence has been placed on active control. A proliferation of other types of diversion structures can be found in Europe and other parts of the world (Fraser, 1966). In populated areas the possibility of avalanche debris being diverted to the benefit of some but the detriment of others must always be considered in their design.

Other protective measures that make a structure more resistant to impact forces may be integrated into existing or proposed facilities; such measures include reinforcement, angled walls and roofs, and an assortment of protective shutters and doors for buildings located in avalanche paths. These adaptations can be more aesthetically pleasing than retarding or diversion structures, and their cost can be more easily amortized over the long term. Although a safe haven may be created, no protection is provided to people and property located outside the structures. The hazard of access into or out of reinforced structures remains unchanged unless diversion devices are also installed.

New structures built in avalanche paths in the United States may have reinforcing features designed into their construction. Some local building codes require design considerations for inhabited buildings in avalanche paths (Mears, 1980), though uniform engineering standards do not exist. Questions may arise concerning appropriate engineering criteria and liability in the event of design failure.

COMMENTS

1.   No system providing accountability and effective channels for information transfer exists for developing and implementing safe procedures and transmitting related technological developments. In the early years of U.S. avalanche control, procedures for technology transfer were developed by a small group involving the USFS, USFS permittees, and the National Ski Patrol System. No such formal system exists now, although an ad hoc committee on weapons use, established in 1989 (Penniman, 1989b), after this report had passed review, represents a step in the right direction. A follow-up meeting was held in Seattle in May 1990 (D. Abromeit, U.S. Forest Service, written communication, 1990).

2.   Improvements are needed in the handling of explosives. A development program is needed to test alternative weapons delivery systems, including other types of surplus artillery. There is still a need for an accurate and reliable short-range weapon with a large supply of ammunition. Accurate inventories of ammunition are needed, and crew training procedures should be reviewed and improved.

3.   A formalized certification procedure should be established, and information and training should be widely available. Present training programs appear to be derived from the original Memorandum of Understanding and involve U.S. Army and USFS instructors. Areas with military weapons have each developed their own weapons training programs, and in most cases have retained crews for long periods of time. This has developed a local expertise that is stable if slightly ingrown. But the loss of crews through attrition or age, and

the need for additional weapons programs, will inevitably require additional training. There appears to be justification for a uniform nationwide weapons training program to include all explosive systems. Such a program might include

    a.   careful development of instruction manuals;

    b.   basic training in weapons handling to persons lacking experience;

    c.   continued education in training and safety for personnel at all levels of experience;

    d.   development of certification standards based on both written tests and weapons handling ability;

    e.   training in procedures for weapons maintenance and ammunition storage and transportation; and

    f.   training in the documentation, location, and disposal of duds.

4.   The problem with duds is important and is increasing in severity, but despite some efforts to find a replacement for military ordnance, adequate solutions have not been developed. Indeed, use of HEPT rounds will likely exacerbate the problem. Alternatives include the development of a new projectile with self-destruct capabilities, increased emphasis on dud location, and more sensitive fusing. Explosive-carrying cable lift systems enable explosive loads to be retrieved if firing has failed (indeed this is compulsory in France; Brugnot, 1987). Therefore, one possible solution to the problem is to replace artillery with cable delivery systems.

5.   Cable delivery systems offer some potential for U.S. industrial entrepreneurship, but developments in the United States substantially lag those in Europe.

6.   The chief problem with structural control of avalanches is cost. The massive structures needed to stabilize deep snow on steep slopes are expensive to construct and must be regularly inspected and repaired. Yet routine maintenance is difficult to fund.

7.   European experience on structural control procedures may be more or less directly transferred to the United States, if proper site evaluation is conducted prior to design and installation.

NOTES

1.   Apparently, the governing statute is 10 USC 4655: "When required for the protection of public money and property, the Secretary of the Army may lend arms and issue ammunition to federal agencies upon request by agency head" (D. Abromeit, U.S. Forest Service, written communication, 1990.) The latest Memorandum of Agreement with the USFS is dated July 1989, affecting 13 ski areas in 7 states.

2.   Memoranda of Agreement exist between the U.S. Army and state government agencies in Alaska (March 1987), California (November 1989), Colorado (October 1987), Washington (February 1989), and Wyoming (June 1989) (D. Abromeit, written communication, 1990).

6
Forecasting Avalanches

Snow avalanche forecasting is the probabilistic assessment of both current and future snow stability. The philosophical purpose behind such forecasting is to provide information about current mountain conditions that helps people to avoid or to minimize exposure to avalanches. Forecasting is a formidable problem, as future stability assessments require consideration of additional loading in the form of anticipated precipitation and changes in snow strength resulting from temperature-controlled processes within the snowpack. Accurate avalanche forecasts are thus highly dependent on accurate weather forecasts, which are intrinsically difficult in mountainous areas.

FORECASTING ORGANIZATIONS

The oldest avalanche forecasting service is in Switzerland; it was established in 1945. This system, still the world standard, is centralized within a portion of the Federal Institute for Snow and Avalanche Research (FISAR) at Weissfluhjoch-Davos (Jaccard, 1986). The relations between meteorology, snowpack information, and avalanche formation are translated into avalanche hazard warnings that are distributed to the public. In addition, the avalanche warning service accumulates a data base of avalanche, snowpack, and weather information; provides information to the public on preventative and operational avalanche protection; and performs accident analyses for courts of law.

A comprehensive view of avalanche conditions requires the daily acquisition and rapid transmission of observations from all Swiss mountain regions to the FISAR. Thus, about 70 observation stations are distributed throughout the Alps at altitudes of 1,000–2,500 m. These stations are operated part time by local people of various occupations who have received special training; their daily reports reach FISAR each morning through the "Meteor" communication system of the Swiss Institute of Meteorology. Avalanche warnings are broadcast at noon several times a week by radio and can be retrieved by telephone; reports are also issued by television and the press.

France operates a centralized avalanche forecasting service via Météorologie Nationale and Centre d'Etudes de la Neige (de Crecy, 1980; LaFeuille et al., 1987). The latter institute has developed systems to enable local agencies to conduct avalanche hazard forecasting (Navarre et al., 1987). A diversified network of about seven forecast centers is maintained in the Alps and the Pyrenees, with each center issuing a forecast for its local area. In contrast, the forecast system in Austria is less centralized, and each province issues its own hazard bulletin (Bauer, 1972; Rink, 1987; Gallagher, 1981).

In Canada avalanche forecasting is carried out by a number of cooperating agencies, predominantly under the Atmospheric Environment Service of Canada (AESC). This agency receives weather data from its own weather offices, public works departments, ski resorts, national parks, hydro companies, and private individuals. Two main area bulletins are issued by AESC, one for the coastal mountains and one for the interior mountains. More localized avalanche forecasting can be obtained from the individual agencies that relay weather data to the AESC, but only the park rangers in Alberta are currently engaged in public forecasting.

These examples provide a comparative background for forecasting activities in the United States, where avalanche forecasting can generally be described as being carried out on both local and regional scales. In some locations forecasters are concerned with individual avalanche paths, perhaps as few as 5 to 50 in number, such as those located within ski areas, recreational land, or along specific sections of highways. Most highways do not have forecasting programs and wait for actual avalanche debris in the highway before issuing closures.

Other forecasters have the responsibility to provide information covering entire regions without making reference to individual avalanche paths. Examples of regional forecasting, which may include thousands of square kilometers of avalanche terrain, include the organized avalanche information centers in Colorado, Utah, and Washington. These centers derive their financial support from various agencies, which in turn receive the benefit of the mountain weather forecasts and avalanche information [U.S. Forest Service (USFS), National Weather Service, National Park Service, state highway departments, ski-area organizations, mountain clubs, etc.]. The USFS administers and supports the Utah center and also administers the Northwest center with considerable financial support from additional organizations; the Colorado center is administered by the state but relies on numerous other organizations for its financial support. Housing for all centers is provided by National Weather Service Forecast Offices. Some centers claim financial problems; an operating center in Alaska was closed for financial reasons in 1986.

The forecasters at the regional centers generally provide weather, snow stability, and avalanche hazard ratings within a given area, with specific information regarding the range of elevation, slope angle, and aspect. This information is produced at least once daily by the centers and is available to the public through recorded messages and through the media. For example, the Colorado Avalanche Information Center (CAIC) operates from November to April and monitors weather, snowpack, and avalanche data at 32 manned sites (22 ski areas; the remainder are highway and backcountry locations); provides twice-daily forecasts to the public via recorded telephone messages; issues avalanche warning bulletins via the National Oceanic and Atmospheric Administration's Colorado Weatherwire and the news media; and provides avalanche information to the public (Williams, 1986).

The CAIC also maintains a computer data set of mountain weather and avalanche events from about 60 sites throughout the Western mountains, continuing the Westwide Avalanche and Mountain Weather Reporting Network originated by the USFS in 1966. The Westwide network uses standardized instrumentation and data collection procedures and provides valuable statistics on avalanche occurrence and associated snowpack and weather conditions for several thousand avalanche paths in the United States. Modeled in some respects after the Swiss system, the Westwide data set was expected to furnish the basic observations necessary for research efforts toward objective avalanche forecasting in the United States (Martinelli, 1973). Research along these lines was in fact initiated (Judson et al., 1980) but was terminated with the closure of avalanche studies at Fort Collins. The network was also expected to provide the care of long-term avalanche occurrence records so essential for national land-use planning and zoning as well as the basic data for a conceived national avalanche warning system (Martinelli, 1973).

Apart from the Westwide network responsibility, the Northwest center functions on a basis similar to Colorado. Emphasis is on highly detailed mountain weather forecasts (Marriott and Moore, 1984; Ferguson et al., 1989). The Utah Avalanche Forecast Center conducts operations over a smaller region, but it services the most concentrated population of winter backcountry use in the country; its avalanche hotlines receive 50,000 calls per season (Tremper and Ream, 1988).

STATE OF THE FORECASTING ART

To evaluate the probability of an avalanche release at some future time, the forecaster must have access to data that describe both the expected meteorological conditions and the anticipated strength conditions of the snow cover. Because useful snow strength data are extremely difficult to obtain, forecasting methods place greatest emphasis on meteorological variables (LaChapelle, 1980; Buser et al., 1985). The collection of weather data has been aided by advances in digital recording systems and by the development of durable sensors adequate for winter use in mountain locations (Gubler, 1984; Marriott and Moore, 1984).

To some limited extent, weather data can be considered to represent the general conditions of the area of perhaps several square kilometers surrounding the measurement site. In the case of snowpack data, this same assumption cannot be made, since snow properties on a shallow, gently sloping, south-facing slope will differ greatly from those of a thick snow cover on a steep north-facing slope only a few meters away (Dexter, 1986; R. L. Armstrong, 1985). While such snow properties as density, temperature, and crystal type can be measured by standardized methods, the correlation between these and representative snow strengths remains elusive, as does the measurement of snowpack strength. Existing models relating weather and snowpack parameters to snowpack strength are quite complex and thus far offer little practical assistance to the operational avalanche forecaster (Dexter, 1986; Judson et al., 1980; Anderson, 1976).

The essence of the problem in avalanche forecasting is to determine the amount of energy required to trigger an avalanche for a given set of strength conditions. The meteorological, snow structure, and snow mechanics data that actually become input variables for specific forecast methods are determined by both the requirements of the technique being used and the ability of the forecaster to obtain the data. Once the required information has been assembled, a detailed decision-making process is undertaken, whether by actual forecasters using conventional methods, by means of a numerical model, or, as is becoming more typical, by both methods.

Widely practiced traditional methods of avalanche forecasting therefore require a blend of inductive logic and deterministic consideration of meteorological and snow physics parameters to reach actual forecast decisions (LaChapelle et al., 1978; LaChapelle, 1980). Conventional forecasting is thus an art based on experience, intuition, and process-oriented reasoning that is difficult to learn, to teach, and to transfer from one region to another. This situation motivates the continuing search for objective computer-based procedures to aid in decision making—efforts currently concentrated in European institutions. Methods being examined include linear regression, multivariate discriminant analysis, time-series modeling involving nonparametric methods and pattern recognition, numerical deterministic modeling, nearest-neighbor methods, and artificial intelligence (Buser et al., 1987; Bakkehoi, 1987; LaFeuille et al., 1987; LaFeuille, 1989; Navarre et al., 1987).

Although the emphasis on numerical and statistical modeling has focused almost exclusively on meteorological variables, the various methods in current use have produced reasonable results (Buser et al., 1985, 1987). Nevertheless, at present such forecasts achieve a score only comparable to the results obtained by conventional or intuitive methods, where the technique relies almost entirely on the experience of the forecaster (Armstrong and Ives, 1976; Fohn et al., 1977; Buser et al., 1987). Computer techniques do not yet relieve the forecaster from making decisions but do provide a tool by which detailed specific information is made available as part of the basis for forecasting decisions. Such models offer promise, but they require an effective long-term data base of snowpack and meteorological information. In this respect the Westwide data network system is of crucial importance, an "invaluable treasure" in the words of Brugnot (1987).

COMMENTS

1. In the United States no uniform policy exists regarding avalanche forecasting. The administration and funding of forecast centers are fragmented; several centers have financial problems and concern for survival; and the Alaskan center was closed due to lack of funds. Because these centers provide a valuable service, an attempt should be made to find funding resources to ensure their continued operation.

2. Development of new forecasting methodologies for avalanche forecasting is now carried out mainly in Europe, where government-derived financial support is available for such activities. Additional funding would be required to enable forecast centers in the United States to develop, adapt, and test new technologies.

3. A data base essential to future computer-based forecasting in the United States is being maintained at a minimum level by the regional centers and by the Westwide data network. Because this data base is of crucial importance, it should at least be maintained and, if possible, upgraded. Additional funding is needed for equipment maintenance, repair, replacement, and modernization.

7
Avalanche Research

INTERNATIONAL PROGRAMS

Research efforts in the United States substantially lag those abroad. In Japan the Institute of Snow and Ice Studies was established at Nagaoka in 1964 as part of the National Research Center for Disaster Prevention (NRCDP), under the Science and Technology Agency (T. Nakamura, personal communication, 1989). The NRCDP also maintains the Shinjo Branch of Snow and Ice Studies. In these laboratories avalanche research is conducted as one of four principal areas of snow research. Topics include impact measurements at instrumented field sites and experimental chutes, studies of glide phenomena, laboratory investigations, computer flow modeling, and automated warning systems (Nakamura et al., 1981, 1987, 1989; Abe et al., 1987; Sato, 1987).

In Sapporo, Japan, is the Institute of Low Temperature Science, established in 1941 to conduct fundamental and applied research on phenomena occurring in low-temperature climates. This institute, with a staff of 90, has gained international recognition for its work in physical and biological fields of cold-region science. Administered by Hokkaido University, the institute consists of 12 sections, 8 of which consider physical or glaciological topics (N. Maeno, Institute of Low Temperature Science, Sapporo, Japan, personal communication, 1986). Avalanche dynamics is also an important topic in the meteorological section (Shimizu et al., 1980; Akitaya, 1980; Maeno et al., 1987, 1989; Nishimura and Maeno, 1987, 1989; Nishimura et al., 1989).

In addition, a snow and ice laboratory is maintained by the Railway Technical Research Institute, and a number of universities conduct research on snow and avalanches; for example, avalanche research is carried out jointly with ground-failure hazards at Niigata University's Research Institute for Hazards in Snowy Areas (Izumi, 1985; Izumi and Kobayashi, 1986).

In Europe, where about 1,200 fatalities occurred as a result of avalanches in the last decade (Valla, 1987), extensive research is performed. The predominant facility is the Swiss Federal Institute for Snow and Avalanche Research (FISAR), a unique mountain laboratory

52

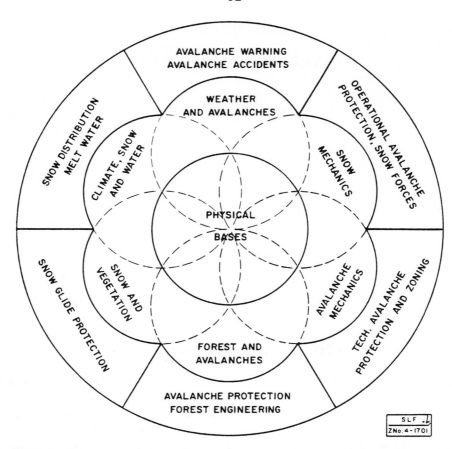

FIGURE 3 Symbolic scheme of research and practice at the Swiss Federal Institute for Snow and Avalanche Research. The overlapping specialized departments meet in the central basic research sphere, and are surrounded by the zone of the practice-related problems.

above the city of Davos (Figure 3). FISAR has been devoted to studying all problems related to snow avalanches for more than 50 years (de Quervain, 1986; Jaccard, 1986; Salm and Gubler, 1987; Gubler and Weilenmann, 1986; Gubler, 1977, 1983, 1985, 1987, 1989; Bachmann, 1987; Buser, 1989; Buser and Good, 1987; Good, 1987; Fohn, 1987). Technology transfer and consultation on avalanche problems are of high priority. FISAR is organized under the Federal Department of the Interior within the Federal Forestry Service and maintains a staff of about 33 in 4 scientific sections: (1) Weather, Snowpack, and Avalanches; (2) Snow and Avalanche Mechanics and Avalanche Constructions; (3) Snow Cover and Vegetation; and (4) Snow and Ice Physics. FISAR operates a 90-station observation network throughout Switzerland, an avalanche warning service, a library, four cold laboratories, instrumented test fields, forestation test fields, an instrumented experimental avalanche chute, both stationary and mobile frequency-modulated continuous wave radars, and mobile Doppler radar units for velocity studies. Operations are funded within the normal budget of the Federal Forestry Service approved by Parliament, an expenditure adequately reimbursed by effective and controlled engineering and increased avalanche safety (Jaccard, 1986). Contributions from the National Foundation for Scientific Research have occasionally been requested.

53

Apart from the Federal Forestry Service, the most direct influence on FISAR operations comes from the 15-member Federal Commission for Snow and Avalanche Research. Since most members are from universities and are competent in fields pertaining to snow research, they act as consultants for the scientific program. Other commission members represent practical aspects, including forestry, civil engineering, transportation, and tourism. Research is also carried out by the Laboratory of Hydraulics, Hydrology, and Glaciology at the Swiss Federal Institute of Technology in Zurich (Hutter and Alts, 1985; Hutter et al., 1987; Hutter and Savage, 1989; Hermann et al., 1987), the University of Bern (Mätzler, 1987), and others.

In Norway substantial avalanche research is carried out at the Norwegian Geotechnical Institute (NGI), Oslo. NGI employs about 180 persons and is supported by government agencies and consulting activities. About 20 percent of its income is government research funding (H. Norem, Norwegian Geotechnical Institute, Oslo, written communication, 1986). The avalanche section of NGI, comprising seven professionals, is responsible for national research on avalanches of all kinds—snow avalanches, rock avalanches and rockfalls, and slush avalanches (cf. Hestnes and Sandersen, 1987; Norem et al., 1987, 1989; Toppe, 1987). The main research projects include field measurements of forces due to snow creep, avalanche speeds and impact pressures, avalanche-produced water waves (NGI, 1984, 1986), and statistical and numerical estimation methods for extreme avalanche runout distances (Bakkehoi, 1987; Lied and Toppe, 1989). A field research station is maintained in Grasdalen in western Norway.

Snow research at NGI benefits from close collaboration with an instrument section and engineers specializing in soil mechanics and rock engineering. NGI has an excellent library and maintains close relations with the Meteorological Institute and the University of Oslo. The mixture of research and consulting activities makes it possible to bring the results of research rapidly into practice; likewise, consulting experience provides useful background for the evaluation of research results (H. Norem, Norwegian Geotechnical Institute, Oslo, written communication, 1986).

Study of physical properties of snow and ice was initiated in the U.S.S.R. in the 1930s (Kuvaeva et al., 1971) and is currently recognized as an independent discipline with as many as 10 scientific organizations working on problems related to snow physics and avalanches. The dominant problem for many of these institutes (e.g., the Alpine Geophysical Institute) is the study of snow cover and avalanches (Glaciological Data, 1984; Sulakvelidze and Dolov, 1969; Kotlyakov et al., 1977; Voitkovskiy, 1987; Zalikhanov et al., 1987), and considerable research has considered such topics as impact dynamics, mathematical flow modeling, snowpack physics, wind-blast effects, and forecasting.

Research in France is carried out at the Division Nivologie CEMAGREF, Centre d'Etudes de la Neige, Saint Martin d'Heres, the Institut de Mécanique de Grenoble, and Association-nationale pour l'Etude de la Neige et Avalanches (ANENA), Grenoble. French research has pioneered such topics as stereophotogrammetric velocity determinations (Brugnot, 1982) and the development of a powder-snow avalanche dynamic code using a blend of theory and modeling in a water-filled channel (Beghin and Brugnot, 1983; Hopfinger, 1983). The most important French achievement in simulation since 1980 considers dense flowing avalanches through explicit solution of Saint-Venant equations (Vila, 1986, 1987). More realistic than previous solution methods (Brugnot and Pochat, 1981), this approach is mathematically complex but is adaptable to such problems as dam-break flooding and the prediction of water waves generated by avalanches (Vila, 1987). [The water wave topic has also been considered by the Université des Sciences et Techniques du Languedoc,

Montpellier (Sabatier, 1986)]. Other topics include snow rheology (Navarre et al., 1987), development of field-based extreme runout criteria (Berthier, 1986), artificial intelligence (LaFeuille et al., 1987; Charlier and Buisson, 1989; Brugnot, 1987), forecast modeling (Navarre et al., 1987), and control devices such as the explosive release system (Borrel, 1987), the DRA avalanche sensor traffic light warning system, snow rakes, and forest protection (Brugnot, 1987). ANENA publishes a journal, Neige et Avalanches, that contains both scientific research and practical studies.

Elsewhere, well-established research investigations are maintained in Austria at the Forsttechnischer Dienst für Wildbach und Lawinenverbauung, in Tirol (Hagen and Hufnagl, 1987; Friedrich, 1987), and at the University of Innsbruck (Lackinger, 1987, 1989; Denoth and Foglar, 1986). Avalanche work is also done in Czechoslovakia (L. Knvazovicky, consultant, Jasna, Czechoslovakia, written communication, 1986); at several institutes in China (e.g., Academia Sinica, Xinjiang, and Lanzhou (Jiaqi and Ruji, 1980; Yanlong et al., 1980); and at universities in Yugoslavia, West Germany, and Iceland (Björnsson, 1980). In Italy avalanche research is performed at the Experimental Centre for Avalanches and Hydrogeological Defence, operating under the Regione Veneto Dipartimento Foreste in Belluno, and at the privately financed Vanni Eigenmann Fondazione Internazionale in Milano, which undertakes avalanche safety and rescue research (Eigenmann, 1978). In addition, there has been recent research in Argentina at the Instituto Argentino de Nivología y Glaciología, supported by CONICET (Argentina's National Research Council) (J. C. Leiva, Instituto Argentino de Nivología y Glaciología, written communication, 1986). Research is also conducted in New Zealand at Otago and Canterbury universities (Fitzharris et al., 1983; Fitzharris and Owens, 1980; Owens and Fitzharris, 1989), with support from the New Zealand Mountain Safety Council in Wellington. India maintains a research center at the foot of the Himalayas (Rao et al., 1987).

Canada, through the National Research Council of Canada, is also actively engaged in avalanche research (McClung, 1987; Schaerer, 1989). Regional research facilities are located at Vancouver and Rogers Pass, British Columbia. Work in Canada covers the full range of snow and avalanche work, including laboratory, field, and theoretical studies. Areas of research concentration include avalanche dynamics, quasistatic and dynamic forces on structures, avalanche prediction, and snow structure (McClung and Schaerer, 1983; McClung, 1977, 1979, 1981; Hungr and McClung, 1987; McClung and Lied, 1987; McClung and Larson, 1989; McClung et al., 1984; Perla, 1978a,b, 1985; Perla et al., 1980; Dozier et al., 1987; Schaerer and Sallway, 1980). Despite the quality of this research, the research group is small and is currently threatened by budget trimming (D. McClung, National Research Council of Canada, written communications, 1987, 1990). Avalanche investigations are considered under the Associate Committee for Geotechnical Research of the National Research Council, which coordinates Canadian research studies concerned with the physical and mechanical properties of the terrain of the dominion. Technical translation of foreign research on avalanches is supported by Canada's National Research Council. The organization also issues the Canadian Avalanche Newsletter and provides headquarters for the Canadian Avalanche Association (McFarlane, 1984).

CURRENT STATUS OF AVALANCHE RESEARCH IN THE UNITED STATES

The level of avalanche research activity in the United States is extremely small compared with federal agency research budgets or the research levels in Europe, Japan, or even Canada.

Avalanche studies are now restricted to a few universities, avalanche forecasting centers, and private individuals who have an interest in avalanches. Such studies are largely unfunded in any formal sense. The Colorado Avalanche Information Center, for example, has entertained the possibility of attempting research in a modest way. However, this would require doubling its small budget of $110,000; this is unrealistic, "since survival, and not expansion, is our major concern" (K. Williams, Colorado Avalanche Information Center, Department of Natural Resources, written communication, 1988). The amount of federal funds that directly support avalanche research is miniscule.

Some support for snow research is provided by federal agencies, but this is not avalanche research. The U.S. Forest Service (USFS) conducts a modest research program involving such topics as blowing snow (Schmidt, 1982, 1986; Schmidt et al., 1984) and snow melting (Kattelmann, 1987; McGurk and Kattelmann, 1986; Bergamon, 1986). But with the demise of the modest but cost-effective USFS avalanche program at Fort Collins, Colorado, support for snow avalanche research has vanished.

The University of Washington was active in avalanche research from 1973 to 1984, with grants from the Federal Highway Administration, the National Science Foundation, and the Washington State Department of Transportation (LaChapelle et al., 1978). In 1985 the Colorado Division of Highways funded the installation of load cells in a reinforced concrete snowshed in the San Juan Mountains (Mears, 1986). Avalanche research at Montana State University was formerly sponsored by the USFS (Lang and Martinelli, 1979a,b; Dent and Lang, 1980). Funds from the Bureau of Reclamation supported a small program of avalanche research (in relation to concern for possible litigation due to its program of cloud seeding) at the University of Colorado in the 1970s and 1980s (R. L. Armstrong, 1988). The University of Washington, Montana State University, the University of California at Santa Barbara, and Colorado State University now conduct funded research on mechanical properties, optical properties, blowing snow, snow melt hydrology, etc., but little if anything in the way of direct avalanche studies.

This lack of avalanche research reflects the absence of organized funding, not the lack of worthwhile research targets. Numerous research areas could improve the technology for forecasting and mitigating avalanche hazards. Ranking high among these are mountain meteorology and the ability to improve wind and snowfall predictions (Speers and Mass, 1986; Rhea, 1978; Dunn, 1983). The majority of avalanches occur during and immediately after storms, so the ability to predict snowfall or snow drift patterns is of primary importance. Considerable research could be devoted to development of meteorological models that utilize large-scale meteorological input from the National Weather Service to improve the accuracy of forecasts for specific mountain ranges. On a smaller scale, such models could perhaps be extended to specific regions, such as recreational areas. These models would necessarily be computer based and would include topographic as well as meteorological factors.

A related need is improving prediction of snow deposition patterns in mountains, given a specific area snowfall, wind speed, and wind direction. Computer-based mathematical models could in principle be developed to allow forecasters to predict deposition patterns in complex terrain (Tesche, 1988), a capability that would also be useful in assessing the effects of modification on snow deposition patterns and snowmelt runoff.

More studies are needed on release mechanisms. Models to predict snowpack strength and density profiles from meteorological data (temperature, solar insolation, wind, snowfall, etc.), for example, represent extremely complicated and perplexing problems that to date have been inadequately addressed (Judson et al., 1980). In the area of snow mechanics, virtually no data exist on strength properties of seasonal snow in the density and grain shape ranges that apply to avalanche conditions or on the spatial distribution of snowpack strength and stress patterns. Post-control release deserves study (Pratt, 1984), as do fracture initiation and propagation (Bader et al., 1989). Similarly, acoustic emissions are of interest as potential indicators of slab instability (St. Lawrence, 1980; Sommerfeld and Gubler, 1983; Watters and Swanson, 1986; McClung, 1987; Leaird and Plehn, 1984).

Other specialized topics concerning materials also need further study since they are essential ingredients in the forecasting process. An example is surface hoar formation and its evolution within the snow cover, for which little quantitative data exist (Colbeck, 1988; Breyfogle, 1986). The ability to predict the precise conditions for surface hoar formation and its growth rate and properties would be useful. Other topics include studies on the formation of wind crusts and sun crusts and their bonding to the overlying snow cover. These special topics, while in themselves not large problems, are important to avalanche prediction and are not currently well understood.

Avalanche dynamics is yet another area in need of thorough investigation, inasmuch as such studies provide basic input for zoning and other types of hazard mitigation. The state of the art has developed to the point where sophisticated computer models could now be developed to investigate avalanche flow over variable terrain for different snow conditions. Topics such as basal friction, turbulence, entrainment, deposition, and three-dimensional effects still need to be better understood, though some progress on these areas has been made (Gubler, 1987, 1989; Norem et al., 1987; Hutter and Alts, 1985; Hutter et al., 1987, Tesche, 1986). The innovative use of radar systems shows promise in dynamic studies of natural avalanches (Gubler and Hiller, 1984; Gubler, 1987). Field measurements of velocity and impact pressure and creep pressure yield information crucial to structural control (Schaerer and Sallway, 1980; Shimizu et al., 1980; Akitaya, 1980; Lang and Brown, 1980; Mears, 1986; Larsen et al., 1985; McClung and Larsen, 1989), though the reliability of some published results is reported to be in question (Brugnot, 1987).

Field and laboratory research is needed to develop and test new methodologies and to refine existing procedures for delineating and mapping avalanche hazards (Martinelli, 1984; Mears, 1984). Additional topics include avalanche control measures, such as reforestation (Montagne et al., 1984; Jaccard, 1985); structural methods and explosive delivery systems (Brugnot, 1987, 1989; Rapin, 1989); and rescue methods, including development of electronic transceivers to locate avalanche victims (Lind and Smythe, 1984; Good, 1986; Dozier et al., 1989).

The social aspects of avalanches and avalanche hazard forecasting have not received much attention and deserve more. The reaction of the recreation industry to forecasting and the manner of preparing and releasing forecasts to achieve maximum acceptance and benefit are several of many social science topics that could be considered.

COMMENTS

1. Avalanche research has been conducted on a small scale at a handful of universities and federal laboratories in the United States, but with the closure of the USFS avalanche program no federal agency currently has a dedicated program to address this hazard. For a consistent national research capability to be established and maintained over the long term, certain programs and divisions of the National Science Foundation (NSF) need to be designated to accept avalanche proposals, and the responsibilities of federal agencies need to be reexamined.

2. The research funding problem is complicated because avalanche research involves a number of separate disciplines. So many facets of engineering and the physical sciences are involved that NSF programs in engineering, mathematical and physical sciences, and atmospheric and earth sciences could all, in principle, entertain proposals on avalanche research. However, since in the past no program in either engineering or the physical sciences clearly accepted responsibility for funding avalanche research proposals, these proposals tended to slip through the cracks in the system. There were valid reasons for this, since (a) NSF programs have a natural preference for concentrating on so-called mainstream research topics, highly visible with respect to program missions; (b) the snow avalanche problem is not as serious a problem as some others and therefore may be placed on a lower priority level; and (c) the avalanche problem has such a broad interdisciplinary nature that, without a concerted effort on the part of funding agencies such as NSF to define appropriate programs responsible for funding, most programs would hesitate to assume responsibility.

To some extent this negative situation may have been ameliorated by the recent reorganization of NSF, in which the Natural and Manmade Hazards Mitigation program was established within the Directorate of Engineering: "The natural hazards of interest to this program are geophysical in nature, and related to the meteorologic, hydrologic and geomorphic extreme events which each year endanger, damage, or destroy lives, property and resources. . . . Research efforts in this program are directed to natural hazards such as hurricanes and tornadoes, floods and droughts, landslides and mudflows, snow drifts and icejams" (NSF, Program Announcement, OMB 3145-0058). This program currently accepts proposals for avalanche research.

Other funding opportunities may exist at NSF. For example, atmospheric sciences is a natural research area for problems such as blowing and drifting snow, cornice formation, snow deposition patterns in mountainous terrain, and precipitation processes. The Experimental Meteorology program in the Division of Astronomical, Atmospheric, Earth, and Ocean Sciences could fund such research. In the Directorate for Engineering, topics such as fluid dynamics, turbulence, and multiphase flow could reasonably be placed under sponsorship of the programs of Engineering Science in Chemical, Biochemical, and Thermal Engineering and Engineering Science in Mechanics, Structures, and Materials Engineering. Other topics, such as avalanche dynamics, avalanche release processes, mechanical properties of snow, and heat and mass transport in snow, could fall within the responsibilities of a number of programs, depending on the particular emphasis given by the principal investigator. No single division can be expected to assume sole responsibility for all avalanche-related problems, since the problems are so strongly interdisciplinary.

Given that the above programs could logically provide support, some means of ensuring that avalanche and snow research proposals have their "day in court" must be implemented.

Unless programs are given official responsibility to include snow, ice, and avalanche topics, such proposals will continue to have difficulty being fairly considered for funding.

As a corrective measure, NSF could clearly identify specific programs as having responsibility for proposals relating to snow and avalanche problems. A mechanism for directing these proposals to the correct program should be instituted. Programs responsible for these types of proposals should be earmarked in the NSF Guide to Programs, so that scientists and engineers can easily obtain some indication of the correct program to which they should submit their proposals. Finally, recognizing that NSF is an organic entity that undergoes periodic restructuring, the problem may require periodic reevaluation.

3. Turning to the question of federal agency involvement, it seems clear that the interdisciplinary nature of snow avalanche studies creates problems analogous to those concerning potential NSF funding. Yet at the same time this diversity may offer flexibility in finding plausible answers to the problem.

Federal agency involvement could assume several possible forms. The most realistic possibilities are the following:

a. Establishment of a national laboratory dedicated to alpine snow and avalanche research. The Swiss Federal Snow and Avalanche Research Institute provides the clear model for such an enterprise.

b. Incorporation of avalanche research into the ongoing research programs of one or several federal agencies. The choice of agency would depend on whether emphasis is placed on materials (avalanches of snow), on processes and hazards (avalanches as a type of slope failure), or on the principal territory affected (avalanches on federal lands).

These lines of thought lead, respectively, to the following possibilities for incorporation of an avalanche research effort into existing agency programs:

a. U.S. Army Cold Regions Research and Engineering Laboratory;

b. U.S. Geological Survey landslide research, as part of a ground-failure hazards-reduction program;

c. USFS's, as part of a mountain snow research program.

Decisions ultimately will be governed by economic and political factors—where can funds be made available for avalanche research, now and in the long term, and in which agency are administrators interested in developing a program of avalanche research? *These are indeed the key questions, for nothing significant will happen unless some individuals step forward to accept the task and a source of funding can be established.*

Perhaps the best and most direct way to establish an avalanche research capability in the United States would be to create a national research center devoted to avalanche problems.[1] Swiss experience indicates that avalanches are indeed a very difficult, complex, and multifaceted phenomenon that can best be studied by research teams at a research center. This would require a budget sufficient for a technical and support staff with the required field, laboratory, and analytical skills. Unfortunately, when weighed against the economic magnitude of the avalanche problem in the United States and the current economic climate for research funding, establishing such a research center would seem hard to justify. Although the federal economic climate could change, and other possibilities for research support could be developed via public endowment, industrial sponsorship, or through such techniques as "snow safety" surcharges attached to commercial ski tickets or backcountry users, at present a national center concept does not seem supportable. Nonetheless, the current situation in which there is no organizational focus for avalanche work and no funding available to support an ongoing program is equally hard to justify. A middle ground should therefore be sought.

If an avalanche program could be incorporated as part of a more general research effort, justification might be more realistic, and the resources of a more diverse group of scientists, engineers, field personnel, and technicians could be utilized. Centering such an effort in a permanent research group seems necessary to assure the long-term records needed for probabilistic solutions.

One possibility involves the U.S. Army Cold Regions Research and Engineering Laboratory (CRREL) at Hanover, New Hampshire, which decades ago briefly supported a review of avalanche research (Mellor, 1978; see also Borland, 1953; Fuchs, 1957). CRREL concentrates its efforts on sea ice, lake and river ice, frozen soils, permafrost, and atmospheric icing. In the area of snow, CRREL provides support for vehicle mobility, material properties, stress wave studies, penetration mechanics, electrical and optical properties, and blowing and drifting snow. Some of this research on material properties has potential application to avalanche technology (Colbeck, 1987), but the applications are indirect. Most current studies have potential military applications in mind. Further, the location of CRREL, in New Hampshire, is not central to U.S. avalanche problems. The solution to the avalanche research question is best sought elsewhere.

Another possibility is to once again incorporate avalanche studies into a mountain snow research program of the USFS. The now-defunct Fort Collins avalanche program began in this fashion, with an overall program including wind transport and deposition, hydrologic aspects of mountain snow cover, and interaction of snow with timber production. The avalanche portion was shut down in 1985, associated with a reduction in hydrologic studies and an increase in acid precipitation research. Such a program could be reinstated.

However, the size of the USFS avalanche research effort in terms of staff and total budget (about $200,000 per annum) made it vulnerable to negative administrative decisions when funds became increasingly difficult to obtain. Furthermore, the interest in mountain snow research was small within the context of the USFS mission, which is focused on the production of timber resources. Other alternatives might provide a relatively more substantial base on which to found and maintain a long-term research effort in avalanches. This is not to suggest that avalanche research should not be carried out by the USFS, for the panel's opinion is that such research would be beneficial. We merely recognize that such a program may be of uncertain longevity, given past experience, and that the research involvement of several agencies can be justified.

As a final possibility the U.S. Geological Survey (USGS) should be considered, since this is the principal federal organization concerned with slope failure (U.S. Geological Survey, 1981, 1982). Public Law 93-288, the Disaster Relief Act of 1974, which includes provisions that the federal government be prepared to issue warnings of disasters to state and local officials and provide them with technical assistance, specifically identifies landslides among the geologic hazards to be addressed. Under this act, the director of the USGS has been delegated specific responsibility for issuing disaster warning "for an earthquake, volcanic eruption, landslide, or other geologic catastrophe."

As a federal agency the USGS embraces those elements of a slope failure program that are of national, overview, or fundamental scientific concern. These elements include research on slope failure processes, with emphasis on mechanics, materials, and rates; prototype and demonstration studies of hazard, risk, and vulnerability assessment; and research on slope failure prediction and the development of model early-warning systems. The USGS also is responsible for positive actions to transfer its research findings to those of federal, state, local, and private groups in whose charge rests hazard-mitigation implementation (USGS Management Implementation Plan, Geologic Hazard Surveys, FY 1986). Within the USGS there is no national center for landslide studies. Instead, such activities are dispersed under the Geologic Division and the Water Resources Division at such locations as Denver, Colorado; Menlo Park, California; Reston, Virginia; and Vancouver, Washington (Cascade Volcano Observatory).

A precedence exists for some snow or ice avalanche research by USGS scientists (Mathes, 1930; Twenhofel et al., 1949; Davis, 1962; Post, 1968; Witkind et al., 1972; Bryant, 1972; Love, 1973; Frank et al., 1975; Luedke, 1976; Plafker and Erickson, 1978; Voight, 1980, 1981; Voight et al., 1981, 1983; Armstrong and Carrara, 1981; Waitt et al., 1983; Waitt, 1990; Pierson et al., 1990; R. Denlinger, U.S. Geological Survey, personal communication, 1986; R. J. Janda, U.S. Geological Survey, personal communications, 1986, 1990; W. Hotchkiss, U.S. Geological Survey, personal communication, 1985; R. L. Christiansen, U.S. Geological Survey, personal communication, 1986). USGS personnel were involved in the Juneau, Alaska, avalanche hazard problem circa 1950 (Twenhofel et al., 1949; R. Miller, communication cited by LaChapelle, 1972) and were instrumental in relocating a school proposed for a hazardous location. Reports in the 1970s reflected regional hazard mapping, whereas most recent studies involve snow-volcano interactions.

The current lack of significant USGS involvement in snow avalanche research reflects several factors, including the perception within the USGS that the topic was authoritatively and comprehensively embraced by the USFS and the inadequacy of funding resources to allow full response to other high-priority slope failure topics such as warning systems and technical assistance responsibilities.

Yet snow avalanche studies are recognized as having direct relevance to landslide research (and vice versa) on processes, hazard delineation, and warning systems. To cite one example, close parallels are recognized between flowing snow and slush avalanches and debris flows (Hestnes and Sandersen, 1987; Nyberg, 1985) and, to cite another, between powder avalanches and turbidity currents (Hermann et al., 1987; Scheiwiller, 1986; Scheiwiller and Hutter, 1983). Methods for delineating and mitigating snow avalanches and other slope failure hazards are similar (Kienholz, 1978; Ives and Bovis, 1978; Mears, 1979; Brabb, 1984; Hansen, 1984; Kockelman, 1986), and research on processes and dynamic simulation originally developed for snow avalanches have been profitably applied to other areas of slope failure research and practice (Lang and Dent, 1983; Trunk et al., 1986).

In principle, and assuming availability of funds, the USGS national landslide program could be strengthened to address the problem of snow avalanches, particularly in areas of process and hazard delineation. This possibility deserves to be explored.

NOTE

1. Previous initiatives in the United States to develop such a national center include the following: (a) a snow and avalanche research and resource center at Fort Collins, Colorado, was proposed as a USFS-founded "Center of Excellence," with cooperation between the USFS and Colorado State University to be carried out under a Memorandum of Understanding (Martinelli, 1978); and (b) a National Avalanche Resource and Research

Center was proposed for the Salt Lake City/Cottonwood Canyon, Utah, location as a USFS-founded entity cooperating with the University of Utah, the U.S. Army Tooele Depot and Dugway Proving Ground, and the National Weather Service (Anderson, 1977). Neither proposal was funded.

8
Problems in Communications

TECHNOLOGY TRANSFER

Essential in fields where research-generated technology is to be implemented, technology transfer is crucial to the operation of avalanche forecasting centers, highway departments, and planning agencies, which must apply new technological information. Currently, there are few mechanisms for technology transfer in the avalanche field. This situation is aggravated by the fact that foreign research and technology developments have outstepped efforts in the United States and by the absence of any federal program with the responsibility to coordinate technological developments. The avalanche situation mimics the deteriorated state of U.S. industrial competitiveness (National Academy of Engineering, 1987).

Technology transfer implies a kind of balanced equation. On one side are laboratory and university-based scientists and engineers—now mainly in institutions abroad—striving to gain a better understanding of the properties and processes of snow and avalanches, while on the other side practitioners attempt to put these results to practical use. Currently, even when research results are published in English, they are generally presented in the language of science or engineering, which is difficult for many practitioners to understand and apply. For instance, new formulations on avalanche flow have been developed that offer a better way to predict hazard boundaries and impact loads on structures. These formulations, claimed to be superior to previous theories, are demonstrated for only a small number of examples. Practitioners may perhaps be expected to be able to follow the reasoning and immediately apply it in their work. However, due to a lack of rigorous scientific training among most practitioners, this simply does not happen.

The Avalanche Review, the official publication of the American Association of Avalanche Professionals, and the International Snow Science Workshop, held in the United States or Canada every 2 years, are useful instruments for technology transfer. Avalanche Review is a nonprofit publication that provides information transfer between researcher and practitioner and background information for the general public. Published six times each year, the journal contains research results written in layperson's terms, book reviews, and news of avalanche incidents and snowpack and weather conditions for the avalanche community in North America and around the world.

In 1986 the American Association of Avalanche Professionals (AAAP) was organized as a nonprofit association with the goals of representing the professional interests of the U.S. avalanche community, contributing to high standards of professional competence, exchanging technical information, acting as a resource base for public awareness programs, and promoting research and development. This fledgling organization has high goals and a small but energetic membership; funds are limited and are derived mainly from membership dues. The Swiss FISAR has offered its cooperation to AAAP in matters concerning the "International Decade for Natural Disaster Reduction" activities (C. Jaccard, Federal Institute for Snow and Avalanche Research, Davos, Switzerland, personal communication, 1986).

Prior to 1984 the U.S. Forest Service (USFS) Alpine Snow and Avalanche Project in Fort Collins, Colorado, served as a repository for avalanche accident data reported on standard avalanche accident forms and published in The Snowy Torrents (Gallagher, 1967; Williams, 1975; Williams and Armstrong, 1984a). These accident data were stored in Colorado State University's mainframe computer system, and programs were written for data analysis. When the USFS transferred administration of the Avalanche Warning Center to the State of Colorado in 1984, this information was entered into a data-base management program on the Colorado Avalanche Information Center's (CAIC) microcomputer. The former Alpine Snow and Avalanche Project collected mountain weather and avalanche event data from numerous ski areas and observation sites in the western states, which are now the basis of the Westwide data network, managed at a reduced level by the CAIC. The CAIC publishes a monthly newsletter, Avalanche Notes, from November through April, that summarizes monthly weather and avalanche events and provides a narrative of avalanche accidents for each month from the western states and Alaska. These data are also stored in Colorado State University's mainframe computer. Support to manage this data base comes from the USFS.

EDUCATION

Information programs are essential for bringing avalanche information to the attention of the public. Any hazard-reduction program depends on public understanding and public support (Kockelman, 1986). Thus, education on avalanche matters, oriented primarily toward those who live, work, or vacation in the mountains, may be undertaken by individuals, agencies, schools, nonprofit organizations, and special-interest groups. Typical techniques are given in the box below.

The need for education is underscored by the fact that in the United States between 1950 and 1985, 75 percent of the 290 known avalanche fatalities were vacationers and of these the majority were traveling in the backcountry (Armstrong and Williams, 1986). In Alaska alone about 260 of the 278 individuals known to have been caught in avalanches between 1980 and 1985 actually triggered the slides that caught them (Fredston and Fesler, 1985). And in a recent survey by the Utah Avalanche Forecast Center of 154 winter backcountry users, respondents each witnessed an average of 5.2 human-triggered avalanches; 4 out of every 10 were themselves caught by avalanches (Tremper and Ream, 1988). The number of avalanche accidents thus continues to climb nationwide as backcountry use increases and more travelers with limited avalanche awareness access mountainous terrain. The high

TYPICAL COMMUNICATION TECHNIQUES FOR AVALANCHE HAZARD REDUCTION

Educational Services

Assisting and cooperating with universities and their extension divisions in the preparation of course outlines, detailed lectures, case books, and display materials.

Contacting speakers and participating as lecturers in regional and community educational programs.

Sponsoring, conducting, and participating in topical and areal seminars, workshops, short courses, technology utilization sessions, cluster meetings, innovative transfer meetings, training symposia, and other discussions with user groups.

Releasing information needed to address critical avalanche hazards early through oral briefings, seminars, map-type "interpretive inventories," open-file reports, reports of cooperating agencies, and "official use only" materials.

Sponsoring or cosponsoring conferences for planners and decision makers at which the result of avalanche studies are displayed and reported to users.

Providing speakers to government, civic, corporate, conservation, and citizen groups and participating in radio and television programs to explain or report on avalanche hazard-reduction programs and products.

Assisting and cooperating with regional and community groups to incorporate avalanche information into school curricula.

Preparing and exhibiting displays that present avalanche information and illustrate their use in hazard reduction.

Attending and participating in meetings with local, district, and state agencies and their governing bodies to present avalanche information.

Guiding field trips to potentially hazardous sites.

Preparing and distributing brochures, films, videotapes, and other visual materials.

Advisory Services

Preparing annotated and indexed bibliographies of avalanche information and providing lists of pertinent reference material to various users.

Assisting local, state, and federal agencies in designing policies, procedures, ordinances, statutes, and regulations that cite or make other use of avalanche information.

Assisting in recruiting, interviewing, and selecting planners, engineers, and scientists by government agencies for which education and training in avalanche information collection, interpretation, and application are criteria.

Assisting local, state, and federal agencies in the design of their avalanche information collection and interpretation programs and in their work specifications.

Providing expert testimony and depositions concerning avalanche research information.

Assisting in the presentation and adoption of plans and plan implementation devices that are based on avalanche information.

Assisting in the incorporation of avalanche information into local, state, and federal studies and plans.

Preparing brief fact sheets or transmittal letters about avalanches explaining their impact on local, state, and federal planning and decision making.

Preparing users in the creation, organization, staffing, and formation of local, state, and federal planning and plan implementation programs so as to assure the proper and timely use of avalanche hazard information.

Preparing and distributing appropriate user guides relating to avalanche processes, mapping, and hazard-reduction techniques.

Preparing model state avalanche safety legislation, regulations, and development policies.

Preparing model local avalanche safety policies, plan criteria, and plan implementation devices.

Review Services

Review of proposed programs for collecting and interpreting avalanche information.

Review of local, state, and federal policies, administrative procedures, and legislative analyses that have a direct effect on avalanche information.

Review of proposed policies, procedures, and legal enactments that cite avalanche information.

Review studies and plans based on avalanche information.

SOURCE: Adapted from Kockelman (1986).

percentage of human-caused accidents and the indication that the same mistakes are made repeatedly point to the urgent need for avalanche education.

A variety of programs and schools offering different levels and types of avalanche training have been established in the United States, including avalanche awareness lectures, intensive avalanche hazard-evaluation workshops (basic and advanced) and professional courses. Avalanche-related courses are offered by only a handful of universities (Montagne, 1980).

Early avalanche training was sponsored by the USFS in the 1950s. In 1971 the USFS founded the National Avalanche School, which is held every 2 years and now has a capacity of more than 200 students per session. Emphasis in the 5-day school is on providing a technical basis for practical work, rather than on state-of-the-art research and high-level technology transfer. Optional Phase II field courses in a number of mountain locations offer the opportunity for site-specific applications of material presented in classroom sessions. Demand for the lecture program has exceeded availability, due to space limitations and the desire for a low student-teacher ratio. Administration of the school was assumed by the National Avalanche Foundation from 1981 to 1985, and in 1986 its administration was again transferred, this time to the National Ski Patrol System (NSPS). Actual instruction in the school is carried out by experts in the avalanche field, and over the years the roster of instructors has not changed much despite shifts in administration.

The NSPS, a nonprofit volunteer winter rescue organization chartered by Congress, is the largest single provider of avalanche education. The NSPS has for many years provided avalanche training to its patrol members and the general public. In a typical year it holds 75 or more basic avalanche courses, each consisting of at least 12 hours of instruction. These courses provide introductory training to over 750 members of the NSPS and at least 400 nonmembers. The National Ski Patrol also conducts about 10 advanced avalanche courses, each consisting of 4 days of classroom and field work. A certificate of completion is awarded annually to about 100 patrollers and 25 nonmembers.

The National Ski Patrol also promotes avalanche awareness through public lectures that each year reach several thousand members of the skiing public. AAAP and NSPS members occasionally make radio and television appearances and provide informational articles for various magazines and newspapers. Such activities are generally undertaken on a volunteer basis but serve a useful purpose, as the public cannot readily interpret avalanche media bulletins without some knowledge of avalanche hazard evaluation.

The American Avalanche Institute (AAI), founded in 1974 by a private individual, was the first private avalanche school in the United States offering both classroom and field training. AAI has held a variety of courses throughout the western states and in New Hampshire, varying from 1 to 4 days in length. Participants are primarily backcountry skiers, climbers, and professional ski patrollers.

The State of Alaska subsidized avalanche education from 1972 through 1986. In 1972 the Department of Natural Resources created the Alaska Avalanche School (AAS), which in 1980 became a primary component of the Alaska Snow Avalanche Safety Program. The AAS conducted more than 100 major workshops involving over 11,000 participant-days of training, provided hundreds of shorter lectures and workshops for schools and civic groups, and generated avalanche information to the public through the media. When funding for the statewide avalanche program was terminated in 1986, the Alaska Mountain Safety Center was established by private individuals as a nonprofit educational organization to operate the Alaska Avalanche School.

Numerous regional and local schools are run by individual search and rescue groups, mountaineering clubs, guiding companies, and recreation equipment stores. While the majority of courses in the United States are geared toward recreationists, there have been a few special courses, such as the AAI's "Avalanche Litigation Workshop" and the AAS's "Avalanche Hazard Evaluation in Land Use Planning" for planners, engineers, and policy-makers.

Education has also been carried out through several other forums; the three regional avalanche forecast centers, for example, have reached hundreds of thousands of people via recorded snow-stability messages as well as short courses and workshops. During the winter of 1985–1986, the Colorado Avalanche Information Center taught 1,184 people, in courses ranging from basic avalanche safety lectures to field sessions for experienced avalanche practitioners (Williams, 1986). However, in recent years avalanche forecast center budgets have been reduced, personnel have been eliminated, and fewer resources have been allotted to avalanche education.

COMMENTS

1. Many foreign nations have their own research centers and have moved ahead of the United States in innovative technologies. The United States would benefit from enhanced access to this increasingly significant body of technological information. Technology transfer could be improved by more frequent seminars, training sessions, and publications to disseminate information on new developments.

2. Technology transfer also remains a problem because of the complex nature of theoretical and practical avalanche technology and the lack of rigorous technical training among most practitioners. Much of the new technology is computer dependent, including such topics as mountain meteorology, blowing snow, avalanche flow, snowpack structural change by metamorphism, and avalanche release mechanisms and mechanics. If this technology is to be successfully transferred to practitioners, appropriate user-friendly software must be developed and documented, demonstrated to the appropriate technicians, and made available with technical support. Even with such software, judgment is likely to remain a problem.

3. Another matter requiring attention is the lack of a centralized repository for snowpack and avalanche information. Individual avalanche forecast centers collect avalanche occurrence and snow stratigraphy data and carry out local investigations, but the data are only incompletely archived and collected for research use.

4. While it is difficult to quantify the success of available educational programs in preventing avalanche accidents, few question their desirability, and several praiseworthy programs have been developed. A basic problem is that funding has been insufficient to sustain some effective programs. Even within the primary target group of recreationists, only a small percentage have received minimal training. Safe, high-quality training is costly, and private schools have found it financially difficult to offer high-quality courses at a reasonable price. The cost of liability insurance has become a major factor in limiting the success of private educational ventures (R. Newcomb, American Avalanche Institute, Wilson, Wyoming, personal communication, 1986; Burr, 1989).

5. A related nationwide problem is that most avalanche instructors are "borrowed" by specific schools from their conventional full-time work. Skilled individuals are often unavailable for training others. Furthermore, knowledgeable individuals are not necessarily

skilled educators. The time and funding needed to train qualified avalanche workers as instructors, or to produce needed educational materials, have always been inadequate. The situation is frustrating because most avalanche accidents are avoidable, and education offers a powerful tool for prevention.

6. The National Avalanche School and comparable AAI programs are basic in nature. More intensive specialized training is needed in blasting, artillery operations, and rescue operations.

9
Conclusions and Recommendations

Snow avalanches are a multifaceted, complex component of the national ground-failure problem (National Research Council, 1985) and the international natural hazard problem (National Research Council, 1987). While snow avalanches do not affect the overall U.S. population as much as other ground failure hazards, they are a problem that requires greater attention and it deserves increased and sustained funding.

Avalanches are the most frequent catastrophic mass movement in the nation and the single greatest natural hazard to winter activities in mountainous areas. Avalanche hazard is becoming more significant as development and recreation increase in mountain regions.

The U.S. scientific and technological effort in avalanche work is minimal, and the nation lags other countries managing this problem. Existing avalanche programs in the United States are small and, on the whole, are declining in response to the withdrawal of previously limited but critical federal funding. There is no national program for avalanche prediction, mitigation, education, or research, nor any formal coordination of these activities at other levels of government. Whereas avalanche management is accorded some local emphasis, current strategies are carried out on an ad hoc reactive basis rather than comprehensively; standardization does not exist from one region to another. Although several agencies are involved in some aspects of avalanche forecasting, including the U.S. Forest Service, the National Weather Service, and the National Park Service, no unifying policy exists.

No federal agency carries out or actively supports avalanche research; develops hazard-delineation or hazard-mitigation methodologies; or provides technical assistance to state, local, and private organizations wishing to reduce avalanche hazards. Research of this kind is carried out by the U.S. Geological Survey for other kinds of slope failures, but the results of such studies have not yet been adapted to slope problems involving snow. Avalanche control methods, especially those involving explosives, present serious hazards both to the public and to operators and can be properly addressed only at the national level.

The following findings and recommendations are presented as a basis for addressing key problems.

NATIONAL LEADERSHIP

There is no overall organization or focus on this increasingly significant natural hazard. As a result, mechanisms for communication, regulation, and support are not well developed. No government agency accepts overall responsibility or takes a leadership position in matters related to avalanche hazard identification, mitigation, relief, or research. Resolution of this issue is of highest priority.

The provisions of the Disaster Relief Act of 1974, as applied to snow avalanches, are not at present adequately addressed: it is one matter to establish such a directive and another for agencies to possess the will, institutional capability, and funding resources for its effective implementation. The International and the United States Decade for Natural Hazard Reduction provide a timely opportunity to focus on these issues, but it remains to be seen whether the opportunity is converted into action.

Recommendations

1. The federal government should assume specific but limited responsibilities for avalanche hazard delineation and control. These could include (a) the development of methodologies for delineation and control on a variety of scales, (b) pilot mapping and control demonstrations, and (c) avalanche mapping and control in support of the missions of federal agencies. In addition, the federal government should work with other parties to provide cooperative support, information, and technical assistance to state, local, and private organizations.

2. Research under national leadership should be undertaken to improve the technical base for avalanche forecasting, control, land-use planning, and public warning systems. A modest program including field, laboratory, and theoretical research on avalanche initiation and dynamics, coupled with avalanche prediction and meteorological models, control measures, and risk appraisal, should be carried out through (a) interdisciplinary research in appropriate federal agencies and (b) support and maintenance of a research capability in universities through funding by the National Science Foundation.

3. To assist, review, and delineate the above tasks within the federal establishment, *the federal government should establish a mechanism for program initiation and coordination among the federal agencies having responsibilities related to slope failure, snow research, administration of federal lands containing avalanche hazards, and administration of forecasting centers.* A short-lived interagency task force or interagency coordinating committee might be an appropriate way to accomplish this important purpose. Agency representatives should include the U.S. Forest Service, U.S. Geological Survey, and National Weather Service, among others.

4. Effective nationwide coordination of avalanche management and research programs is necessary. The coordination entity should not be a federal agency but rather a U.S. national-level committee consisting of representatives from government, academia, industry, and professional organizations. Whatever its nature, the specific interests of federal, state, or local agencies as well as private institutions having responsibility for various aspects of avalanche mitigation should be represented.

5. The purpose of the committee would be to provide sustained momentum and guidance toward the solution of these problems. Such a committee, analogous to the advisory Swiss Federal Commission for Snow and Avalanche Research, could be organized and maintained over the long term under a committee of the National Research Council

charged with reduction of natural hazards or, alternatively, as a panel within the Committee on Glaciology of the NRC's Polar Research Board. (The former may seem preferable inasmuch as avalanches are not strictly a "polar" problem, but the latter may offer advantages of long-term stability).

6. This committee could provide a central focus now absent and offer guidance within the following tasks:

a. Provide sustained authoritative support for federal programs.

b. Provide guidelines for appropriate areas of research at universities and in the private sector.

c. Establish and coordinate a program of technology transfer that will closely monitor the extensive avalanche work being done in other parts of the world and make it available to domestic research and application communities. This will provide a cost-effective method for maintaining a state-of-the-art expertise in the United States.

d. Establish a centralized information archive to manage the wide range of technical data and educational materials pertinent to avalanche work.

e. Provide a forum to encourage legislative innovation.

HAZARD DELINEATION AND REGULATION

One of the most effective ways to reduce avalanche damage is to locate development only on low-hazard ground and to dedicate high-hazard ground to open space and low-intensity use. Where land values are high, expensive engineering solutions may be justified. Land-use control programs are best carried out at the local level, but they require adequate mapping and enabling legislation that may involve state or federal entities.

The development and implementation of design and building practices that minimize avalanche damage are to some extent complicated by geographic (climate zone) variations on the nature of avalanche risk, the small number of trained geotechnical engineers assigned to code development and enforcement, and the lack of national leadership. Greater emphasis could be placed on the application of current knowledge as a basis for code development.

Recommendations

1. The federal government should encourage the consideration and effective use of land-use controls by state and local governments to mitigate avalanche hazards.

2. States should mandate, enable, or otherwise provide encouragement to local governments to adopt regulations that will lead to the identification of avalanche hazards and to their avoidance through the control of land development.

3. Local governments should require developers to map and disclose information about hazardous areas.

4. Local governments should post readily visible warning signs to alert prospective developers and purchasers to an avalanche hazard. Such warnings should be based on adequate data and be posted where avalanche areas intersect or abut public rights-of-way, such as "slide area" signs along highways. Warnings can also take the form of rubber-stamped notations on subdivision plots or on building or zoning permits.

5. The technical base for code development should be maintained by technology transfer. The codes themselves should be developed at state and local levels in response to regional and local conditions.

CONTROL MEASURES

Much avalanche incidence and damage can be reduced by prudent and innovative structural control measures. Critical issues involving public and user safety arise in the case of explosive control measures, particularly with regard to the use of artillery weapons.

Recommendations

1.   Pilot studies of structural control effectiveness should be conducted in a variety of settings to establish the adequacy of design criteria and to identify appropriate practices in terms of costs and benefits.

2.   The results should be applied at state and local levels in response to regional and local conditions, with technology transfer made available through federal agencies.

3.   A federal program should be established to ensure the safe use of explosive systems, including military artillery, in avalanche control programs. This program should consider such issues as safety training and certification standards for users, inventory of critical munitions, spare parts, aging ammunition, ammunition storage and transportation, and the problem of lost and unexploded but fully armed shells (duds).

4.   Many of the operational problems associated with artillery control are eliminated by use of cable delivery systems. Further attention to cable delivery technology is therefore encouraged.

FORECASTING

Forecasting provides information about current mountain conditions that helps people to avoid or to minimize exposure to avalanches. Despite the valuable service provided by regional forecast centers in the United States, the administration and funding of these centers are fragmented, and several centers have financial problems and concern for survival. Development of new forecasting methodologies for avalanche forecasting is now carried out mainly in Europe, where government financial support is available.

Recommendations

1.   Regional avalanche hazard forecasting centers should receive adequate federal assistance, in compliance with the directive to issue warnings in the Disaster Relief Act of 1974.

2.   A data base essential to future computer-based forecasting in the United States is being maintained at minimum levels by the regional centers and by the Westwide data network. This data base should at least be maintained and, if possible, upgraded. Additional federal funding is needed now for equipment maintenance, repair, replacement, and modernization.

RESEARCH

While there are many practical working models of avalanche initiation and dynamic behavior, quantitative understanding of the process is limited. Related topics include snow mechanics, mountain meteorology, and flow modeling, which have direct impact on forecasting and the avoidance and control of avalanche hazards. Research is also needed to develop and test new control methods for reinforcing the snowpack, for releasing subcritical snow slabs by safe explosive procedures, and for designing structures to resist avalanche damage. New developments in geophysics such as acoustics, frequency modulated, continuous wave (FMCW) and Doppler radar, and satellite microwave radiometry are being developed and have potential application in avalanche hazard reduction. Automated data collection systems based on electronic instrumentation are available for remote measurement of snow conditions, movement, and meteorology. Such systems could serve the dual roles of monitoring and early warning.

Recommendations

1. Existing avalanche forecasting centers should be funded at a higher level to allow a modest program of research that would include gathering information for a national data base on avalanches. Such information is essential for statistically based forecast procedures.
2. Additional funding sources for individual project research should be clearly designated to encourage university participation. Funding should be made available for addressing the key issues of the avalanche problem.
3. An ideal solution would be the establishment of a national research center dedicated to avalanche research.
4. A more cost-effective alternative would be to attach avalanche research to an existing laboratory or center dedicated to a relevant but broader-based program, as presently accomplished by the Norwegian Geotechnical Institute. The U.S. Geological Survey Landslide Program is suggested as a prime candidate. Other candidates include the regional experiment stations of the U.S. Forest Service and the U.S. Army Cold Regions Research Laboratory in New Hampshire.
5. Research carried out by state, local, or private entities should be encouraged, particularly in regard to control measures and field sites.
6. Cooperative investigations between U.S. and foreign research institutions should be encouraged.

COMMUNICATIONS

There is currently inadequate transfer and dissemination of existing and new technologies and of stored data applicable to avalanche identification, analysis, and control. With research support approaching zero in the United States, the information gap triggered by the U.S. Forest Service's withdrawal from technology transfer has not been filled. Activities of the American Association of Avalanche Professionals, the International Snow Science Workshop, and the National Ski Patrol System help reduce this information gap as well as the incidence of recreational accidents, but all are in need of additional support.

Recommendations

1. Existing information dissemination programs should be supported with increased federal and state funding.
2. Partnerships and cost-sharing enterprises between public and private sector special interest groups should be encouraged.

3.   Programs should be established to translate significant foreign research findings for wider use in the United States and to publish and disseminate key technical documents.

CONCLUDING REMARKS

Snow avalanche risk is increasing measurably in the United States as development and recreational use of mountain areas accelerate. Despite the destructive nature of snow avalanches and the dangers they pose for mountain residents and tourists, there is no coordinated national leadership in avalanche hazard management. There is no national program to set policy, define standards and guidelines, or establish effective communication in such critical areas as prediction, education, land-use planning, and basic research. No government agency assumes responsibility for coordination of the existing ad hoc efforts of government and the private sector.

There is a void of leadership, and individuals and groups involved in the identification, evaluation, and solution of problems related to avalanche hazard no longer have a specific agency or facility to consult for guidance and expertise.

The cost-effective solutions proposed here include the establishment of a national-level committee representing government, academia, and industry to encourage, coordinate, and assist the federal government in assuming its specific but limited responsibilities for hazard delineation and mitigation and modest agency support for research and communication programs aimed at hazard mitigation.

References

Abe, O., T. Nakamura, T. E. Lang, and T. Ohnuma. 1987. Comparison of simulated runout distances of snow avalanches with those of actually observed events in Japan. International Association of Hydrological Sciences Publications 162:463-473.

Abromeit, D. 1988. Military weapons for avalanche control. The Avalanche Review 6(4):4-5.

Akifyeva, K. V., et al. 1978. Avalanches of the U.S.S.R. and their influences on the formation of natural territory complexes. Arctic and Alpine Research 10(2):223-233.

Akitaya, E. 1980. Observations of ground avalanches with a videotape recorder (VTR). Journal of Glaciology 26(94):493-496.

Ambach, W., and F. Howorka. 1965. Avalanche activity and free water content of snow at Obergurgl. International Association of Hydrological Sciences Publications 69:65-72.

Anderson, D. 1977. Proposal to the U.S. Forest Service for establishment of an Office of Avalanche Control National Coordinator and National Avalanche Resource Center. National Ski Areas Association Avalanche Committee. May 24.

Anderson, E. A. 1976. A point energy and mass balance model of a snow cover. NOAA Technical Report NWS 19. Silver Spring, Maryland: U.S. Department of Commerce, Office of Hydrology.

Armstrong, B. 1976. Century of struggle against snow: a history of avalanche hazard in San Juan County, Colorado. Colorado University Institute of Arctic and Alpine Research Occasional Paper No. 18.

Armstrong, B. 1979. Highway safety plan, Cottonwood Canyon, Wasatch National Forest, Salt Lake County, Utah. (with B. Von Allmen, D. Anderson, C. G. Blake, L. Fitzgerald, R. Linquist, B. Sandahl, and R. Thomas).

Armstrong, B. 1980. A quantitative analysis of avalanche hazard on U.S. Highway 550, southwestern Colorado. Pp. 95-104 in Proceedings of Western Snow Conference, St. George, Utah, April 14-16.

Armstrong, B. R., and P. Carrara. 1981. Avalanche hazard areas in the Telluride Mining District, Colorado. U.S. Geological Survey Map I-1316. Scale 1:24,000.

Armstrong, B., and K. Williams. 1986. The Avalanche Book. Golden, Colorado: Fulcrum, Inc.

Armstrong, R. L. 1977. Continuous monitoring of metamorphic changes of internal snow structure as a tool in avalanche studies. Journal of Glaciology 19(81):325-334.

Armstrong, R. L. 1981. Some observations on snow cover temperature patterns. Technical Memorandum No. 133. Ottawa, Canada: National Research Council of Canada.

Armstrong, R. L. 1985. Metamorphism in a subfreezing, seasonal snow cover: the role of thermal and vapor pressure conditions. Ph.D. thesis, Department of Geography and Cooperative Institute for Research in Environmental Sciences (CIRES), University of Colorado.

Armstrong, R. L. 1988. Snow and avalanche research in the San Juan Mountains, southwestern Colorado. 1971-1987. Final report to the Bureau of Reclamation, Denver, Colorado (in preparation).

Armstrong, R. L., and B. R. Armstrong. 1987. Snow and avalanche climates of the western United States. International Association of Hydrological Sciences Publications 162:281-294.

Armstrong, R. L., and J. D. Ives. 1976. Avalanche release and snow characteristics, San Juan Mountains, Colorado. INSTAAR Occasional Paper No. 19. Boulder, Colorado: University of Colorado.

Atwater, M. M. 1968. The Avalanche Hunters. Philadelphia, Pennsylvania: MacRai Smith Co.

Austin, C. F., M. Osburn, C. Halsey, and C. Wilson. 1974. Premature detonation studies with selected explosive materials for avalanche control. Naval Weapons Center Report TS 74-219. China Lake, California: Detonation Physics Division.

Avalanche Review. 1988. Artificial or natural? The Avalanche Review 6(5):2.

Bachmann, O. 1987. Energy loss of snow blocks passing through a supporting structure. International Association of Hydrological Sciences Publications 162:613-622.

Bader, H., H. Gubler, and B. Salm. 1989. Modeling initial fracture and fracture propagation causing slab avalanche releases. Proceedings of the International Glaciological Symposium, Lom, Norway.

Bailey, R. A., P. R. Beauchemin, F. P. Kapinos, and D. W. Klick. 1983. The volcano hazards program: objectives and long-range plans. U.S. Geological Survey Open File Report 83-400.

Bakkehoi, S. 1987. Snow avalanche prediction using a probabilistic method. International Association of Hydrological Sciences Publications 162:549-550.

Bauer, B. 1972. Avalanches as natural hazards in Austria. In Proceedings of the 22nd International Geographical Congress, Calgary, Alberta, July 24-30.

Beghin, P., and G. Brugnot. 1983. Contribution of theoretical and experimental results to powder-snow avalanche dynamics. Cold Region Science and Technology 8:63-67.

Bergamon, J. A. 1986. Electrical measurements of snow wetness in undisturbed snow. Pp. 64-68 in Proceedings of the International Snow Science Workshop, Tahoe, California.

Berthier, B. 1986. Evaluation statistique des limits maximals atteintes par les avalanches à partir de données topographiques (à paraître).

Björnsson, H. 1980. Avalanche activity in Iceland, climatic conditions, and terrain features. Journal of Glaciology 26:13-23.

Bleuer, H. 1989. Blasting of hanging glacier ice. Proceedings of the International Snow Science Workshop, Whisler, British Columbia.

Borland, W. M. 1953. Investigation of snow conditions causing avalanches. U.S. Army Corps of Engineers Interim Report.

Borrel, G. 1987. Evolution récente des méthodes de déclenchement artificiel en France. International Association of Hydrological Sciences Publications 162:623-626.

Brabb, E. E. 1984. Innovative approaches to landslide hazard and risk mapping. Pp. 307-323 in Proceedings of the 4th International Symposium on Landslides, Toronto, Canada, Vol. 1.

Breyfogle, S. R. 1986. Growth characteristics of hoarfrost with respect to avalanche occurrence. Pp. 216-222 in Proceedings of the International Snow Science Workshop, Tahoe, California.

Brown, J. J. 1982. Role of the U.S. Forest Service at Mount St. Helens. Special Publication 63:206. California Department of Conservation, Division of Mines and Geology.

Brown, R. L. 1980. Propagation of stress waves in alpine snow. Journal of Glaciology 26(94):235-244.

Brown v. MacPherson's, Inc. 1975. Washington 545, P. 2d 13. Supreme Court of Washington.

Bryant, B. 1972. Map showing avalanche areas in the Aspen Quadrangle, Pitkin County, Colorado. U.S. Geological Survey Map I-785.

Brugnot, G. 1982. La stéréophotogrammetrie à cadence rapide d'avalanches. Mitteilugen der Forstlichen Bundes versuchsanstalt, 144, Heft.

Brugnot, G. 1987. Avalanche zoning, dynamics and control. Recent work done in France. International Association of Hydrological Sciences Publications 162:521-536.

Brugnot, G. 1989. CATEX: a recent set of rules. Proceedings of the International Glaciological Symposium, Lom, Norway.

Brugnot, G., and R. Pochat. 1981. Numerical simulation study of avalanches. Journal of Glaciology 27(95):77-88.

Bundesant für Forstwesen. FISAR. 1984. Richlinien zur Berucksichtigung der Lawinengefahr bei raumwirksamen Tatigkeiten. Bern u. Davos.

Burr, E. 1989. Problems with liability insurance at Liberty Bell Heliskiing. In Proceedings of the International Snow Science Workshop, Whisler, British Columbia.

Buser, O. 1989. Two years experience of operational avalanche forecast using the nearest neighbors method. In Proceedings of the International Glaciological Symposium, Lom, Norway.

Buser, O., and W. Good. 1987. Acoustic, geometric, and mechanical parameters of snow. International Association of Hydrological Sciences Publications 162:61-71.

Buser, O., P. Fohn, W. Good, H. Gubler, and B. Salm. 1985. Different methods for the assessment of avalanche danger. Cold Regions Science and Technology 10(3):199-218.

Buser, O., M. Buetler, and W. Good. 1987. Avalanche forecasting by the nearest neighbor method. International Association of Hydrological Sciences Publications 162:557-570.

Butler, D. R. 1986. Spatial and temporal aspects of the snow avalanche hazard, Glacier National Park, Montana. Pp. 223-230 in Proceedings of the International Snow Science Workshop, Tahoe, California.

Cazabat, C. 1972. Les cartes de localisation probable des avalanches. Paris: L'Institut Géographique Nationale, XII Congrès Internationale de Photogrammetrie à Ottawa, Ontario, July 23-August 4.

Charlier, C., and L. Buisson. 1989. Avalanche path analysis with knowledge based system. In Proceedings of the International Glaciological Symposium, Lom, Norway.

Clark, S. L. 1988. Mountain hazards mapping: a critical review. Unpublished master's thesis, Department of Geography, University of Colorado.

Colbeck, S. C. 1980. Thermodynamics of snow metamorphism due to variations in curvature. Journal of Glaciology 26(4)291-301.

Colbeck, S. C. 1982. An overview of seasonal snow metamorphism. Review of Geophysics and Space Physics 20(1):45-61.

Colbeck, S. C. 1987. A review of the metamorphism and classification of seasonal snow in cover crystals. International Association of Hydrological Sciences Publications 162:3-34.

Colbeck, S. C. 1988. On the micrometeorology of surface hoar growth on snow in mountainous areas. Boundary Layer Meteorology 44:1-12.

Conway, H., and J. Abrahamson. 1988. Snow slope stability—a probabilistic approach. Journal of Glaciology 34(118):170-177.

de Crecy, L. 1980. Avalanche zoning in France: regulation and technical bases. Journal of Glaciology 26(94):325-330.

de Quervain, M. 1987. 50 years of snow and avalanche research on the Weissfluhjoch. Mitteilungen des Eidgenossischen Institut für Schnee-und Lawinenforschung. 44(January).

Davis, G. H. 1962. Erosional features of snow avalanches, Middle Fork Kings River, California. U.S. Geological Survey Professional Paper 450-D:D122-125.

Délégation aux Risques Majeurs. Mai 1985. Plan d'exposition au risque d'avalanche. Catalogue de sesures de prévention. Edition provisiore.

Denoth, A., and A. Folgar. 1986. Recent developments of snow moisture dialectric devices. Pp. 72-76 in Proceedings of the International Snow Science Workshop, Tahoe, California.

Dent, J. D., and T. E. Lang. 1980. Modeling of snow flow. Journal of Glaciology 26(94): 1311-40.

Dexter, L. 1986. Aspect and elevation effects on the structure of the seasonal snow cover in Colorado. Unpublished Ph.D. thesis, Department of Geography, University of Colorado.

Dombroski, R. 1988. Bridger Bowl air blasting. The Avalanche Review 6(4):5.

Dozier, J., R. E. Davis, and R. Perla. 1987. On the objective analysis of snow microstructure. International Association of Hydrological Sciences Publications 162:49-60.

Dozier, J., R. Faisant, L. Heywood, and G. Reitman. 1989. Comparison of avalanche beacons at 2275 Hz and 475 kHz. In Proceedings of the International Snow Sciences Workshop, Whisler, British Columbia. In press.

Dunn, L. B. 1983. Quantitative and spatial distribution of winter precipitation along Utah's Wasatch Front. NOAA Technical Memorandum NWS WR-181.

Eigenmann, R. 1978. Means for location of avalanche victims in the past and future. In Avalanche Control, Forecasting, and Safety. Technical Memorandum No. 120:284-293. National Research Council of Canada, Associate Committee on Geotechnical Research.

Fagan, J. E., and D. Cortum. 1986. Avalanche control: negligence over strict liability. University of San Francisco Law Review 20(4):719-738.

Ferguson, S. A. 1984a. Strength comparisons between avalanche and non-avalanche snowpacks. Pp. 124-128 in Proceedings of the International Snow Science Workshop, Aspen, Colorado.

Ferguson, S. A. 1984b. The role of snowpack structure in avalanching. Ph.D. Dissertation, University of Washington.

Ferguson, S. A., R. Marriott, M. Moore, and P. Hayes. 1989. Avalanche weather forecasting at the Northwest Avalanche Center, Seattle, Washington. In Proceedings of the International Snow Science Workshop, Whisler, British Columbia. In press.

Fitzharris, B. B. 1981. Frequency and climatology of major avalanches at Rogers Pass, 1909-1977. DBR Paper No. 956. National Research Council, Canada Association Committee on Geotechnical Research.

Fitzharris, B. B., and I. F. Owens. 1980. Avalanche atlas of the Milford Road and an assessment of the hazard to traffic. New Zealand Mountain Safety Council. Avalanche Committee Report No. 4.

Fitzharris, B. B., D. McNulty, I. F. Owens, and I. D. Miller. 1983. Pilot avalanche forecasting project for the Craigieburn Range, New Zealand. Weather and Climate 4:52-58.

Fohn, P. M. B. 1987. The stability index and various triggering mechanisms. International Association of Hydrological Sciences Publications 162:195-214.

Fohn, P., W. Good, P. Bois, and C. Obled. 1977. Evaluation and comparison of statistical and conventional methods of forecasting avalanche hazards. Journal of Glaciology 19(81):375-387.

Frank, D., A. Post, and J. D. Friedman. 1975. Recurrent geothermally induced debris avalanches on Boulder Glacier, Mount Baker, Washington. Journal of Research, U.S. Geological Survey 3(1):77-87.

Fraser, C. 1966. The Avalanche Enigma. Chicago, Illinois: Rand-McNally.

Fredston, J., and D. Fesler. 1985. Snow sense—a guide to evaluating avalanche hazard. Anchorage: Alaska Department of Natural Resources.

Freer, G. L., and P. A. Schaerer. 1980. Snow avalanche hazard zoning in British Columbia, Canada. Journal of Glaciology 26(94):345-354.

Frick, F. C. 1985. The contribution of skiing to the Colorado economy—1985 update: Denver, Colorado: Browne, Bortz, and Coddington, Inc., 8 pp.; cf. F. Frick, D. Coddington, and H. Rubenstein. 1982. The contribution of skiing to the Colorado economy, cited in Utah Law Review 1985(4):814; C. Steffan. 1984. The impact of skiing on the Utah economy. Bureau of Economic and Business Research, University of Utah.

Friedl, J. 1974. Kippel: A Changing Village in the Alps. New York: Holt, Rinehart, and Winston.

Friedrich, D. 1987. Die Praxis der Sprengfundierung in der Lawinenverbauung. Mitteilungen des Eidgenossischen Institut für Schnee-und Lawinenforschung, 43:7-17.

Frutiger, H. 1964. Snow avalanches along Colorado mountain highways. U.S. Forest Service Research Paper RM-7.

Frutiger, H. 1970. Der Lawinenzoneplan (LZP), Schweizerische Zeitschrift für Forstwesen, 121, Jahrbuch, Nr. 4, pp. 246-276. (Translation No. 11, Alta Avalanche Study Center).

Frutiger, H. 1972. Report of the findings of the avalanche hazard inventory. Pp. 53-90 in Geophysics Hazards Investigations for the City and Borough of Juneau, Alaska. Technical Supplement.

Frutiger, H. 1980. History and actual state of legalization of avalanche zoning in Switzerland. Journal of Glaciology 26(94):313-324.

Fuchs, A. 1957. Effective use of explosives on snow. U.S. Snow, Ice, and Permafrost Research Establishment. Special Report 23.

Gallagher, D. G., ed. 1967. The Snowy Torrents: Avalanche Accidents in the United States 1910-1966. Alta Avalanche Study Center, U.S. Forest Service.

78

Gallagher, D. G. 1981. Information and warning programs for back-country travellers. Proceedings of Avalanche Workshop held in Vancouver, British Columbia, 3-5 November 1980. Technical Memorandum 133:176-189. Ottawa: National Research Council of Canada.

Garbolino, J. D. 1986. *Hahn and Nelson* v. *Alpine Meadows Ski Corp., et al.* Ruling regarding strict liability contentions on motion for non-suit. Pp. 183-188 in Proceedings of the International Snow Science Workshop, Tahoe, California.

Gerasimov, I. P., and Zvonkova, T. B. 1974. Natural hazards in the territory of the U.S.S.R.: study, control and warning. Pp. 243-251 in Natural Hazards: Local, National, Global, G. F. White, ed. New York: Oxford University Press.

Gerdes, S. 1988. Avalanches and legal liability. The Avalanche Review 7(1):2.

Glaciological Data. 1984. Soviet avalanche research. Avalanche bibliography update: 1977-1983. Report GD-16, World Data Center for Glaciology (Snow and Ice). Boulder, Colorado: CIRES.

Gmoser, H. 1978. Dealing with avalanche problems in helicopter skiing. Pp. 252-259 in Avalanche Control, Forecasting, and Safety. Technical Memorandum 120. National Research Council of Canada, Associate Committee on Geotechnical Research.

Good, W. 1986. Electronic tranceivers for locating avalanche victims, an optimal strategy for the primary search. Pp. 177-182 in Proceedings of the International Snow Science Workshop, Tahoe, California.

Good, W. 1987. Thin sections, serial cuts, and 3-D analysis of snow. International Association of Hydrological Sciences Publications 162:35-47.

Ground Failure. 1985. What causes landslides? Ground Failure (National Research Council) 2:17-19.

Gubler, H. 1977. Artificial release of avalanches by explosives. Journal of Glaciology 19(81):19-429.

Gubler, H. 1983. Kuenstliche Ausloesung von Lawinen durch Sprengungen. Mitteilungen des Eidgenossischen Institut für Schnee-und Lawininforschung. No. 36. Davos/Weissfluhjoch, Switzerland.

Gubler, H. 1984. Remote instrumentation for avalanche warning systems and snow cover monitoring. Pp. 137-142 in Proceedings of the International Snow Science Workshop, Aspen, Colorado.

Gubler, H. 1985. Model for dry snow metamorphism by interparticle vapor flux. Journal of Geophysical Research 90(D5):8081-8092.

Gubler, H. 1987. Measurements and modelling of snow avalanche speeds. International Association of Hydrological Sciences Publications 162:405-420.

Gubler, H. 1988. Comparison of different avalanche dynamic models. Annals of Glaciology.

Gubler, H. 1989. Comparison of different avalanche dynamic models. In Proceedings of the International Glaciological Symposium, Lom, Norway.

Gubler, H., and M. Hiller. 1984. The use of microwave FMCW radar in snow and avalanche research. Cold Regions Science and Technology 9:109-119.

Gubler, H., and P. Weilenmann. 1986. Seasonal snow cover monitoring using FMCW radar. Pp. 87-97 in Proceedings of the International Snow Science Workshop, Tahoe, California.

Hackett, S. W., and D. Fesler. 1980. Informal cooperative state-federal avalanche-warning system and public education program for south-central Alaska, U.S.A. Journal of Glaciology 26(94):497-500.

Hackett, S. W., and H. S. Santeford. 1980. Avalanche zoning in Alaska, U.S.A. Journal of Glaciology 26(94):377-392.

Hackman, R. J. 1965. Interpretation of Alaskan post-earthquake photographs. Photogrammetric Engineering 31:604-610.

Hagen, G., and H. Hufnagl. 1987. Beobachtungen an Katastrophenlawinen der Jahre 1981, 1984, and 1985 soure Ergebnisse von Nachrechnungen der Lawinenwirkungen. Mitteilungen des Eidgenossischen Institut für Schnee-und Lawinwnforschung 43:7-17.

Hansen, A. 1984. Landslide hazard analysis. Pp. 523-602 in Slope Instability, D. Brunsden and D. B. Prior, eds. New York: John Wiley & Sons.

Hart, K. 1972. Report on the Behrends Avenue avalanche path. Pp. 95-148 in Geophysical hazards investigation for the city and borough of Juneau, Alaska. Technical Supplement.

Hermann, F., J. Hermann, and K. Hutter. 1987. Laboratory experiments on the dynamics of powder snow avalanches. International Association of Hydrological Sciences Publications 162:431-440.

Hestnes, E. 1985. A contribution to the prediction of slush avalanches. Annals of Glaciology 6:1-4.

Hestnes, E., and K. Lied. 1980. Natural hazard maps for land-use planning in Norway. Journal of Glaciology 26(94):331-343.

Hestnes, E., and F. Sandersen. 1987. Slushflow activity in the Rena district, north Norway. International Association of Hydrological Sciences Publications 162(31):7-330.

Hoagland, J. 1988. John Herbert. The Avalanche Review 6(5):1.

Hopfinger, E. J. 1983. Snow avalanche motion and related phenomena. Annual Review of Fluid Mechanics 15:47-76.

Hungr, O., and D. M. McClung. 1987. An equation for calculating snow avalanche run-up against barriers. International Association of Hydrological Sciences Publications 162:605-612.

Hutter, K., and T. Alts. 1985. Ice and snow mechanics, a challenge to theoretical and applied mechanics. In Theoretical and Applied Mechanics, F. I. Niordson and N. Olhoff, eds. Elsevier Science Publications 163:217.

Hutter, K., and S. Savage. 1989. A new computational model for flow avalanches. In Proceedings of the International Glaciological Symposium, Lom, Norway.

Hutter, K., F. Szidarovszky, and S. Yakowitz. 1987. Granular shear flows as models for snow avalanches. International Association of Hydrological Sciences Publications 162:381-394.

Ives, J. D., and M. J. Bovis. 1978. Natural hazards maps for land-use planning, San Juan Mountains, Colorado, U.S.A. Arctic and Alpine Research. 10(2):185-212.

Ives, J. D., and P. B. Krebs. 1978. Natural hazards research and land-use planning responses in mountainous terrain: the town of Vail, Colorado, Rocky Mountains, U.S.A. Arctic and Alpine Research 10(2):213-222.

Ives, J. D., and M. Plam. 1980. Avalanche hazard mapping and zoning problems in the Rocky Mountains, with examples from Colorado, U.S.A. Journal of Glaciology 26(94):363-375.

Izumi, K. 1985. Mobility of large-scale avalanche. Annual Report Saigai-ken (The Research Institute for Hazards in Snowy Areas). Niigata University 7:187.

Izumi, K., and S. Kobayashi. 1986. The movement of powder snow avalanche as recorded on a seismograph. Annual Report Saigai-ken (The Research Institute for Hazards in Snowy Areas) Niigata University 8:99-104.

Jaccard, C. 1985. Snow and avalanches in forests, a Swiss research program. Pp. 11-18 in Proceedings of the IUFRO Meeting and Study Tour—Mountain Forests, Snow and Avalanches.

Jaccard, C. 1986. The Federal Institute for Snow and Avalanche Research at the Weissfluhjoch/Davos. FISAR Special Publication, Davos (revised and translated version of report in German by M. de Quervain).

Jahns, R. H. 1978. Landslides. Pp. 58-65 in Geophysical Predictions—Studies in Geophysics. Washington, D.C.: National Academy of Sciences.

James, R. C. 1981. Alaska avalanche forecast system. In Proceedings of Avalanche Workshop, Vancouver, British Columbia, November 3-5, 1980. Technical Memorandum 133:205-214. Ottawa: National Research Council.

Jiaqui, Q., and H. Ruji. 1980. The avalanches of December 1966 in Western Tien Shan, China. Journal of Glaciology 26(94):512-514.

Jochim, C. L., et al. 1988. Colorado landslide hazard mitigation plan. Colorado Geological Survey Bulletin 48.

Johnson, J. B. 1978. Stress waves in snow. Ph.D. dissertation, University of Washington.

Johnson, J. B. 1980. A model for snow-slab failure under conditions of dynamic loading. Journal of Glaciology 26(94):245-254.

Judson, A., C. F. Leaf, and G. E. Brink. 1980. Process-oriented model for simulating avalanche danger. Journal of Glaciology 26(94):53-63.

Kalatowski, M. 1988. The avalanche history of Alta. (Including "the Binx Sandahl Years" by D. Abromeit). The Avalanche Review 7(3):1-11.

Kattelmann, R. 1984. Wet slab instability. Pp. 102-108 in Proceedings of the International Snow Science Workshop.

Kattelmann, R. 1987. Some measurements of water movement and storage in snow. International Association of Hydrological Sciences Publications 162:245-254.

Kienholz, H. 1978. Maps of geomorphology and natural hazards of Grindewald, Switzerland. Scale 1:10,000. Arctic and Alpine Research 10(2):169-184.

Kockelman, W. J. 1986. Some techniques for reducing landslide hazards. Bulletin of Association of Engineering Geology 23:29-52.

Kotlyakov, V. M., B. N. Rzhevskiy, and V. A. Samoylov. 1977. The dynamics of avalanching in the Khibins. Journal of Glaciology 19(81):431.

Kristensen, K. 1986. Snow avalanche damage in Norway 1985/86. Norges Geotekniske Institutt Report 58810-10.

Kuvaeva, G. M., G. K. Sulakvelidze, V. S. Chitadze, L. S. Chotorlishvili, and A. M. El'Mesov. 1971. Physical properties of the snow cover of the Greater Caucasus (translated from Russian). Indian National Scientific Documentation Centre, New Delhi, with support from USDA-FS and NSF, TT 71-51039.

LaChapelle, E. R. 1956. Alta administrative report for 1955/56. U.S. Forest Service, Utah Avalanche Forecast Center Historical File.

LaChapelle, E. R. 1962. Recent progress in North American avalanche forecasting and control. Journal of Glaciology 679-685.

LaChapelle, E. R. 1966. Avalanche forecasting—a modern synthesis. International Association of Hydrological Sciences Publications 69:350-356. Gentburgge, Belgium.

LaChapelle, E. R. 1968. The character of snow avalanching induced by the Alaska earthquake. Pp. 355-361 in The Great Alaska Earthquake of 1964. Part A: Hydrology. Publication 1603. Washington, D.C.: National Academy of Sciences.

LaChapelle, E. R. 1972. Report on the Behrends Avenue avalanche and other avalanche hazards in the Greater Juneau Borough. Appendix VI in Geophysical Hazards Investigation for the city and borough of Juneau, Alaska. Technical Supplement. Pp. 149-172.

LaChapelle, E. R. 1977. Alternate methods for the artificial release of snow avalanches. Journal of Glaciology 19(81):389.

LaChapelle, E. R. 1978. New experiments for methods of avalanche releases. Pp. 56-60 in Avalanche Control, Forecasting, and Safety. National Research Council of Canada. Associate Committee on Geotechnical Research. Technical Memorandum No. 120.

LaChapelle, E. R. 1980. Fundamental processes in conventional avalanche forecasting. Journal of Glaciology 26(94):75-84.

LaChapelle, E. R. 1981. Report on a review of the Alaska avalanche and fire weather forecast. U.S. Forest Service. Unpublished document.

LaChapelle, E. R. 1985. The ABC of Avalanche Safety. The Mountaineers. 2nd Ed. Seattle.

LaChapelle, E. R., S. A. Ferguson, R. T. Marriott, M. B. Moore, F. W. Reanier, E. M. Sackett, and P. L. Taylor. 1978. Central avalanche hazard forecasting: summary of scientific investigations. Research Project Y-1700. Washington State Transportation Department Research Program Report No 23.4.

Lackinger, B. 1987. Stability and fracture of the snow pack for glide avalanches. International Association of Hydrological Sciences Publications 162:229-241.

Lackinger, B. 1989. Supporting forces and stability of snow slab avalanches. In Proceedings of the International Glaciological Symposium, Lom, Norway.

LaFeuille, J. 1989. Computer aided avalanche forecasting in France. In Proceedings of the International Snow Science Workshop, Whisler, British Columbia.

LaFeuille, J., P. David, J. Konig-Barde, E. Pahaut, and C. Sergent. 1987. Intelligence artificielle et prévision de risques d'avalanches. International Association of Hydrological Sciences Publications 162:537-547.

Lang, T. E., and R. L. Brown. 1980. Numerical simulation of snow avalanche impact on structures. USDA Forest Service Research Paper RM-216.

Lang, T. E., and J. D. Dent. 1983. Evaluation of the fluid dynamic properties of mudflows on Mount St. Helens. Final report prepared for U.S. Department of the Interior, Bureau of Reclamation, Washington, D.C.

Lang, T. E., and M. Martinelli. 1979a. Application of numerical transient fluid dynamics to snow avalanche flow. Part I. Development of computer program AVALNCH. Journal of Glaciology 22:107-115.

Lang, T. E., and M. Martinelli. 1979b. Application of numerical transient fluid dynamics to snow avalanche flow. Part II. Avalanche modeling and parameter error evaluation. Journal of Glaciology 22:117-126.

Larsen, J. O., D. M. McClung, and S. B. Hansen. 1985. The temporal and spatial variation of snow pressure on structures. Canadian Geotechnical Journal 22(2):166-171.

Lazard, A. J. 1986. Available defense systems in the starting zone. Pp. 77-78 in Proceedings of the International Snow Science Workshop, Tahoe, California.

Leaird, J. D., and M. Plehn. 1984. Acoustic emission monitoring in avalanche prone slopes. Pp. 451-466 in Proceedings of the 3rd Conference on Acoustic Emission/Microseismic Activity in Geologic Structures and Materials, Translated Technical Publication. Clausthal, Germany.

Lied, K., and R. Toppe. 1989. Calculation of maximum snow avalanche runout distance based on topographic parameters identified by digital terrain models. In Proceedings of the International Glaciological Symposium, Lom, Norway.

Lind, D. A., and W. R. Smythe. 1984. Avalanche beacons—working principles, specifications, and comparative properties. Pp. 48-53 in Proceedings of the International Snow Science Workshop, Aspen, Colorado.

Love, J. D. 1973. Map showing snowslide possibilities of the Jackson quadrangle, Teton County, Wyoming. U.S. Geological Survey. Miscellaneous Investigation Series I-769-D. Scale 1:24,000.

Luedke, R. G. 1976. Map showing potential snow avalanche areas in the Telluride Quadrangle, San Miguel, Ouray, and San Juan counties. U.S. Geological Survey Map MF-819.

McCarty, D., R. L. Brown, and J. Montagne. 1986. Cornices: their growth, properties, and control. Pp. 41-45 in Proceedings of the International Snow Science Workshop, Tahoe, California.

McClung, D. M. 1977. Direct simple shear tests on snow and their relation to slab avalanche formation. Journal of Glaciology 19(81):101-109.

McClung, D. M. 1979. Shear fracture precipitated by strain softening as a mechanism of dry slab avalanche release. Journal of Geophysical Research 84(B7):3519-3526.

McClung, D. M. 1981. Fracture mechanical models of dry slab avalanche release. Journal of Geophysical Research 86(B11):10783-10790.

McClung, D. M. 1987. Mechanics of snow slab failure from a geotechnical perspective. International Association of Hydrological Sciences Publications 162:475-508.

McClung, D. M., and J. Larsen. 1989. Effect of structure boundary conditions on snow creep pressures. In Proceedings of the International Glaciological Symposium, Lom, Norway.

McClung, D. M., and K. Lied. 1987. Statistical and geometrical definition of snow avalanche runout. Cold Regions Science and Technology 13(2):107-119.

McClung, D. M., and P. A. Schaerer. 1981. Snow avalanche size classification. National Research Council of Canada. Technical Memorandum No. 133, pp.12-30.

McClung, D. M., and P. A. Schaerer. 1983. Determination of avalanche dynamic friction coefficients from measured speeds. Annals of Glaciology 4:170-173.

McClung, D. M., J. O. Larsen, and S. B. Hansen. 1984. Comparison of snow pressure measurements and theoretical predictions. Canadian Geotechnical Journal 21(2):250-258.

McGurk, B. J., and R. C. Kattelmann. 1986. Water flow rates, porosity, and permeability in snowpacks in the central Sierra Nevada. Pp. 359-366 in Proceedings of the American Water Resources Association Symposium on Cold Regions Hydrology, Fairbanks, Alaska.

McFarlane, R. C. 1984. Institutional arrangements for snow avalanche management in Canada. Pp. 84-89 in Proceedings of the International Snow Science Workshop, Aspen, Colorado.

McFarlane, R. C. 1986. Avalanche hazard and climate in Baxter State Park, Maine. Pp. 6-12 in Proceedings of the International Snow Science Workshop, Tahoe, California.

Maeno, N., R. Naruse, and K. Nishimura. 1987. Physical characteristics of snow avalanche debris. International Association of Hydrological Sciences Publications 162:421-427.

Maeno, N., et al. 1989. Dynamical characteristics of a large-scale powder-snow avalanche. In Proceedings of the International Glaciological Symposium, Lom, Norway.

Marbouty, D. 1980. An experimental study of temperature-gradient metamorphism. Journal of Glaciology 26(94):303-312.

Marler, C., and A. L. Fink. 1986. An avalauncher fitting system. Pp. 79-80 in Proceedings of the International Snow Science Workshop, Tahoe, California.

Marriott, R. T., and M. B. Moore. 1984. Weather and snow observations for avalanche forecasting: an evaluation of errors in measurement and interpretation. Pp. 143-154 in Proceedings of the International Snow Science Workshop, Aspen, Colorado.

Martinelli, M., Jr. 1973. Snow avalanches—a recommended program for the U.S. Forest Service in the 1970's. Internal Document U.S. Forest Service, November 3.

Martinelli, M., Jr. 1984. The Goat Lick avalanches of 1979 and 1982. Pp. 198-207 in Proceedings of the International Snow Science Workshop, Aspen, Colorado.

Mathes, F. E. 1930. Geologic History of the Yosemite Valley. U.S. Geological Survey Professional Paper 160.

Mätzler, C. 1987. Microwave sensors for measuring avalanche critical snow parameters. International Association of Hydrological Sciences Publications 162:149-160.

Mears, A. I. 1976. Guidelines and methods for detailed snow avalanche hazard investigations in Colorado. Colorado Geological Survey Bulletin 38.

Mears, A. I. 1979. Colorado snow avalanche area studies and guidelines for avalanche-hazard planning. Special Publication 7. Denver, Colorado: Colorado Geological Survey.

Mears, A. I. 1980. Municipal avalanche zoning: contrasting policies of four western United States communities. Journal of Glaciology 26(94):355-362.

Mears, A. I. 1981. Design criteria for avalanche control structures in the runout zone. USDA-Forest Service, Rocky Mountain Forest and Range Experimental Station. General Technical Report RM-84.

Mears, A. I. 1984. Climate effects on snow avalanche travel distances. Pp. 80-83 in Proceedings of the International Snow Science Workshop, Aspen, Colorado.

Mears, A. I. 1986. Instrumentation of avalanche loads, East Riverside avalanche path, Colorado. Pp. 108-110 in Proceedings of the International Snow Science Workshop, Tahoe, California.

Mellor, M. 1978. Dynamics of snow avalanches. Pp. 753-792 in Rockslides and Avalanches, 1, Natural Phenomena, B. Voight, ed. Amsterdam, Netherlands: Elsevier Scientific Publishing Co.

Montagne, J. 1980. The university course in snow dynamics—a stepping stone to career interests in avalanche hazards. Journal of Glaciology 26(94):97-104.

Montagne, C., J. Montagne, T. Rayne, and A. Satterlee. 1984. New developments for control of snow avalanches in Western European Alps. Pp. 30-35 in Proceedings of the International Snow Science Conference, Aspen, Colorado.

Moskalev, Y. D. 1966. On the mechanism of the formation of wet snow avalanches. International Association of Hydrological Sciences Publications 69:196-198.

NGI. 1984. Statens Naturskadefond Ardalstangen, Ardal Kommune, Vurdering av fare for flodbølge ved skred I kleppura. Norges Geotekniske Institutt 83483-1. October 12.

NGI. 1986. Bolgeoppskylling mot byen maarmorilik pa Grønland som følge av eventuelle fjellskred fra fjellet "den sorte Engel." Norges Geotekniske Institutt 85418-1. November 18.

Nakamura, H., O. Abe, and T. Nakamura. 1981. Impact of snow blocks sliding down from roofs against walls (in Japanese). Natural Research Center for Disaster Prevention, Japan. Report 25.

Nakamura, T., H. Nakamura, O. Abe, A. Sato, and N. Numano. 1987. A newly designed chute for snow avalanche experiments. International Association of Hydrological Sciences Publications 162:441-451.

Nakamura, T. et al. 1989. Blowing and drifting snow observed on the Tsugaru-plain in winter 1986-1987. In Proceedings of the International Glaciological Symposium, Lom, Norway.

National Academy of Engineering. 1987. Strengthening U.S. Engineering Through International Cooperation: Some Recommendations for Action. Washington, D.C.: National Research Council.

National Research Council. 1985. Reducing Losses from Landsliding in the United States. Washington, D.C.: National Academy Press.

National Research Council. 1987. Confronting Natural Disasters: An International Decade for Hazard Reduction. Washington, D.C.: National Academy Press.

National Research Council. In press. Mitigating Losses from Land Subsidence in the United States. Washington, D.C.: National Academy Press.

Navarre, J. P., A. Taillefer, E. Flavigny, J. Desrues, and T. Gauthier. 1987. Mécanique de la neige. Essais en laboratoire sur la résistance de la neige. International Association of Hydrological Sciences Publications 162:129-137.

Niemczyk, K. 1984. Factors comprising county/municipal land-use controls addressing snow avalanches. Pp. 90-94 in Proceedings of the International Snow Science Conference, Aspen, Colorado.

Nishimura, K., and N. Maeno. 1987. Experiments on snow-avalanche dynamics. International Association of Hydrological Sciences Publications 162:395-404.

Nishimura, K., and N. Maeno. 1989. Contribution of viscous force to the avalanche dynamics. In Proceedings of the International Glaciological Symposium, Lom, Norway.

Nishimura, K. et al. 1989. Internal structures of large-scale powder-snow avalanches. In Proceedings of the International Glaciological Symposium, Lom, Norway.

Nobles, L. H. 1965. Slush avalanches of northern Greenland and classification of rapid mass movement. International Association Hydrological Sciences Publications 69:267-272.

Norem, H. 1978. Use of snow fences to reduce avalanche hazards. Pp. 67-72 in Avalanche Control, Forecasting, and Safety. National Research Council of Canada. Associate Committee on Geotechnical Research. Technical Memorandum No. 120.

Norem, H., F. Irgens, and B. Schieldrop. 1987. A continuum model for calculating snow avalanche velocities. International Association of Hydrological Sciences Publications 162:363-379.

Norem, H., et al. 1989. Simulation of snow avalanche flow in run-out zones. In Proceedings of the International Glaciological Symposium, Lom, Norway.

Nyberg, R. 1985. Debris flow and slush avalanches in northern Swedish Lappland. Medd. Lunds Universitats Geografiska Inst. Avh. XCVII, 1-122.

Olshansky, R. B., and J. D. Rogers. 1987. Unstable ground: landslide policy in the United States. Ecology Law Quarterly 13(4):939-1006.

Onesti, L. J. 1985. Meteorological conditions that initiate slush flows on the Central Brooks Range, Alaska. Annals of Glaciology 6:23-25.

Onesti, L. J. 1987. Slush flow release mechanism: a first approximation. International Association of Hydrological Sciences Publications 162:331-336.

82

O'Riordan, T. 1974. The New Zealand natural hazard insurance scheme: application to North America. Pp. 217-219 in Natural Hazards: Local, National, Global, G. F. White, ed. New York: Oxford University Press.

Owens, I., and B. B. Fitzharris. 1989. Assessing avalanche risk levels in Fiordland, New Zealand. In Proceedings of the International Glaciological Symposium, Lom, Norway.

Paine, J., and L. Bruch. 1986. Avalanche of snow from roofs of buildings. Pp. 138-142 in Proceedings of the International Snow Science Workshop, Tahoe, California.

Penniman, D. 1986. The Alpine Meadows avalanche trial: conflicting viewpoints of the expert witnesses. Pp. 189-194 in Proceedings of the International Snow Science Workshop, Tahoe, California.

Penniman, D. 1987. The Alpine Meadows avalanche trial: conflicting view points of the expert witnesses. International Association of Hydrological Sciences Publications 162:665-667.

Penniman, D. 1989a. The political dilemma of avalanche hazard zoning. The Avalanche Review 7(4):8.

Penniman, D. 1989b. Solving the dilemma of weapons use for avalanche control. The Avalanche Review 8(1):8.

Perla, R. I. 1970. Snow avalanches of the Wasatch front. Utah Geological Association Publication 1.

Perla, R. I. 1978a. Failure of snow slopes. Pp. 731-752 in Rockslides and Avalanches, 1, Natural Phenomena, B. Voight, ed. Amsterdam, Netherlands: Elsevier Scientific Publishing Co.

Perla, R. I. 1978b. High explosives and artillery in avalanche control. Pp. 42-49 in Avalanche Control, Forecasting, and Safety. National Research Council of Canada. Associate Committee on Geotechnical Research. Technical Memorandum No. 120.

Perla, R. I. 1980. Avalanche release, motion, and impact. Pp. 397-462 in Dynamics of Snow and Ice Masses, S. C. Colbeck, ed. New York: Academic Press.

Perla, R. I. 1985. Snow in strong or weak temperature gradients, part II, section plane analysis. Cold Regions Science and Technology 11:181-186.

Perla, R. I., and K. Everts. 1983. On the placement and mass of avalanche explosives: experience with helicopter bombing and pre-planted charges. Annals of Glaciology 4:222-227.

Perla, R. I. and M. Martinelli, Jr. 1976. Avalanche Handbook. 2nd Revised Edition (1978). USDA Forest Service, Agriculture Handbook 489.

Perla, R. I., and C. S. L. Ommanney. 1985. Snow in strong or weak temperature gradients. Part 1: experiments and qualitative observations. Cold Regions Science and Technology 11(1):23-36.

Perla, R. I., T. T. Cheng, and D. M. McClung. 1980. A two-parameter model of snow-avalanche motion. Journal of Glaciology 26(94):197-207.

Pierson, T. C., R. J. Janda, J. C. Thouret, and C. A. Borrero. 1990. Origin, flow behavior, and deposition of eruption-triggered lahars on November 13, 1985, Nevado del Ruiz, Colombia. Journal of Volcanology and Geothermal Research.

Plafker, G., and G. E. Ericksen. 1978. Nevada Huascaran avalanches, Peru. Pp. 277-314 in Rockslides and Avalanches, 1, Natural Phenomena, B. Voight, ed. Amsterdam, Netherlands: Elsevier Scientific Publishing Co.

Post, A. 1968. Effects on glaciers. In The Great Alaskan Earthquake of 1964. Part A: Hydrology. Publication 1603:266-308. Washington, D.C.: National Academy of Sciences.

Pratt, T. 1984. Snow creep as a model for post-control releases. Pp. 58-66 in Proceedings of the International Snow Science Workshop, Aspen, Colorado.

Ramsli, G. 1974. Avalanche problems in Norway. Pp. 175-180 in Natural Hazards: Local, National, Global, G. F. White, ed. New York: Oxford University Press.

Rao, N. M., N. Rangachary, V. Kumar, and A. Verdhen. 1987. Some aspects of snow cover development and avalanche formation in the Indian Himalaya. International Association of Hydrological Sciences Publications 162:453-462.

Rapin, F. 1989. CATEX reliability. In Proceedings of the International Glaciological Symposium, Lom, Norway.

Rapp, A. 1960. Recent development of mountain slopes in Karkevagge and surrounding northern Scandinavia. Geografiska Annaler 42(2-3):65-200.

Ream, D. 1990. Avalauncher standard operational procedure. The Avalanche Review 8(5):8.

Rhea, J. O. 1978. Orographic precipitation model for hydrometeorological use. Ph.D. dissertation, Colorado State University.

Rink, C. 1987. Doctrine of geosystems and statistical methods as a means of avalanche forecast. International Association of Hydrological Sciences Publications 162:581-582.

Roch, A. 1949. Report on snow and avalanche conditions in the U.S.A. Western ski resorts, from the 26th of January to the 24th of April, 1949. Report No. 174. Davos, Switzerland: Swiss Federal Institute for Snow and Avalanche Research.

Roethlisberger, H. 1978. Eislawinwen und Ausbruche von Gletscherseen. Jahrbuch der Schweizerischen Naturforschenden Gesellschaft, Wissenschaftilicher Teil:170-212.

Rold, J. W. 1979. Avalanche hazard planning in Colorado; the state and local government relationships. Paper presented at the Symposium on Snow in Motion, Ft. Collins, Colorado, August 15.

Sabatiere, P. C. 1986. Formation of waves by ground motion. Pp. 725-759 in Encyclopedia of Fluid Mechanics. Houston, Texas: Gulf Publishing Co.

St. Lawrence, W. F. 1980. The acoustic emission response of snow. Journal of Glaciology 26(94):204-216.

Salm, B., and H. Gubler. 1987. Measurements and analysis of the motion of dense flow avalanches. Annals of Glaciology 6:26-34.

Sato, A. 1987. Velocity of plastic waves in snow. International Association of Hydrological Sciences Publications 162:119-128.

Schaerer, P. 1989. Avalanche hazard index. In Proceedings of the International Glaciological Symposium, Lom, Norway.

Schaerer, P. A., and A. A. Sallaway. 1980. Seismic and impact-pressure monitoring of flowing avalanches. Journal of Glaciology 26(94):179-187.

Scheiwiller, T. 1986. Dynamics of powder snow avalanches. Mitteilung Nv. 81 der Versuchanstalt für Wasserbau, Hydrologie, Glaziologie der Eidg. Technischen Hochschule, Zurich.

Scheiwiller, T., and K. Hutter. 1983. On shear flow of cohesionless granular materials down an inclined chute. In Advances in the Mechanics and the Flow of Granular Materials, vol. 2, M. Shahinpoor, ed. Houston, Texas: Gulf Publishing Co.

Schmidt, R. A. 1982. Vertical profiles of wind speed, snow concentration, and humidity in blowing snow. Boundary Layer Meteorology 23:223-246.

Schmidt, R. A. 1986. Snow surface strength and the efficiency of relocation of snow. Pp. 355-358 in Proceedings of the Cold Regions Hydrology Symposium. Bethesda, Maryland: American Water Resources Association.

Schmidt, R. A., R. Meister, and H. Gubler. 1984. Comparison of snow drifting measurements at an alpine ridge crest. Cold Regions Science and Technology 9:131-141.

Schuster, R. L., and R. W. Fleming. 1986. Economic losses and fatalities due to landslides. Bulletin of Associated Engineering Geologists 23(1):11-28.

Seligman, G. 1962. Snow Structures and Ski Fields. Edinburgh: R & R Clark Ltd. [Originally published in 1936 by MacMillan and Co., Ltd., London]

Sherretz, L. A., and W. Loehr. 1983. A simulation of the costs of removing snow from county highways in Colorado. Report of Weather Modification Program Colorado Department of Natural Resources. U.S. Bureau of Reclamation Cooperative Agreement No. 1-07-81-VO226.

Shimizu, H., et al. 1980. A study of high-speed avalanches in the Kurobe Canyon, Japan. Journal of Glaciology 26(94):141-151.

Slingerland, R. L., and B. Voight. 1979. Occurrences, properties, and predictive models of landslide-generated water waves. Pp. 317-397 in Rockslides and Avalanches, 2, Engineering sites, B. Voight, ed. Amsterdam, Netherlands: Elsevier Scientific Publishing Co.

Sommerfeld, R. A. 1983. Branch grain theory of temperature gradient metamorphism. Journal of Geophysical Research 88(2):1484-1494.

Sommerfeld, R. A., and H. Gubler. 1983. Snow avalanches and acoustic emissions. Annals of Glaciology 4:271-276.

Sommerfeld, R. A., and R. M. King. 1979. Recommendation for the application of the Roche index for slab avalanche release. Journal of Glaciology 22(88):547-549.

Speers, P. and C. Mass. 1986. Diagnosis and prediction of precipitation in regions of complex terrain. Report No. WA-RD-91.1. Washington State Department of Transportation.

Spray, R. 1983. Safety of the avalauncher. San Dimas Testing Program, U.S. Forest Service Region 3 Recreation File 2300, January 7.

Starr, C. 1969. Social benefits versus technological risk; what is our society willing to pay for safety? Science 165(3899):1232-1238.

Sulakvelidze, G. K., and M. A. Dolov. 1969. Physics of snow, avalanches, and glaciers (translated from Russian). Indian National Scientific Documentation Centre, New Delhi, with support from U.S. Department of Agriculture/Forest Service and National Science Foundation. Publication TT69-53051.

Taylor, D. A. 1985. Snow loads on sloping rock: two pilot studies in the Ottawa area. Canadian Journal of Civil Engineering 12(3):334-343.

Tesche, T. W. 1977. Avalanche zoning: current status, obstacles and future needs. In Proceedings of the 45th Annual Western Snow Conference, Albuquerque, New Mexico, April 18-21.

Tesche, T. W. 1986. A three-dimensional dynamic model of turbulent avalanche flow. Pp. 111-137 in Proceedings of the International Snow Science Workshop, Tahoe, California.

Tesche, T. W. 1988. Numerical simulation of snow transport, deposition, and redistribution. In Proceedings of the Western Snow Conference.

Thomman, R. A. 1986. Research and development pertaining to steel wire rope net systems for the prevention of snow avalanches. Pp. 201-206 in Proceedings of the International Snow Science Workshop, Tahoe, California.

Toppe, R. 1987. Terrain models—a tool for natural hazard mapping. International Association of Hydrological Sciences Publications 162:629-638.

Tremper, B. 1990. GAZ.EX: the continuing saga of avalanche control technology. The Avalanche Review 8(5):5.

Tremper, B., and D. Ream. 1988. Utah backcountry user survey: The Avalanche Review 7(1):1-4.

Trunk, F. J., J. D. Dent, and T. E. Lang. 1986. Computer modeling of large rock slides. American Society of Civil Engineers Journal of Geotechnical Engineering 112:348-360.

Twenhofel, W. S., et al. 1949. Letter report to the Superintendent of Schools City of Juneau, in relation to proposed school in Block F of Highlands Subdivision. File report, City of Juneau (see also LaChapelle, 1972, p. 172).

UGMS. 1983. Governor's conference on geologic hazards. Utah Geological and Mineral Survey. Utah Department of Natural Resources Circular 74.

UNESCO. 1971. Avalanche Atlas. International Commission on Snow and Ice of the International Association of Hydrological Sciences. Paris, France: UNESCO.

U.S. Department of Agriculture/Forest Service. 1971. The training program in winter sports administration. U.S. Forest Service Task Force Report.

U.S. Department of Agriculture/Forest Service. 1983. The principal laws relating to Forest Service activities. USDA/FS Agriculture Handbook No. 453. Washington, D.C.: U.S. Government Printing Office.

U.S. Geological Survey. 1977. Warning and preparedness for geologic-related hazards. Federal Register 42(70):19292-19296.

U.S. Geological Survey. 1981. Goals, strategies, priorities and tasks of a national landslide hazard-reduction program. U.S. Geological Survey Open File Report 81-987.

U.S. Geological Survey. 1982. Goals and tasks of the landslide part of a ground-failure hazards-reduction program. U.S. Geological Survey Circular 880.

Valla, F. 1987. Accidents d'avalanches dans les Alpes au cours de la décennie 1975-1985. International Association of Hydrological Sciences Publications 162:647-652.

Varnes, D. J. 1978. Slope movement types and processes. Pp. 11-33 in Landslides—Analysis and Control, R. L. Schuster and R. J. Krizek, eds. Transportation Research Board Special Report 176. Washington, D.C.: National Academy of Sciences.

Vila, O. P. 1987. La prévision des vagues produites par la chute d'une avalanche dans une retenue. International Association of Hydrological Sciences Publications 162:509-578.

Vila, S. P. 1986. Sur la théorie et l'approximation numérique des problèmes hyperboliques non linéares. Application aux équations de Saint-Venant et à la modelization des avalanches. (These Paris VII).

Voight, B., ed. 1978. Rockslides and Avalanches 1: Natural Phenomena. Amsterdam, Netherlands: Elsevier Scientific Publishing Co.

Voight, B. 1980. Slope stability hazards: Mount St. Helens Volcano, Washington. Vancouver, Washington. U.S. Geological Survey File Report (May 1, 1980).

Voight, B. 1981. Time scale for the first moments of the May 18 eruption. U.S. Geological Survey Professional Paper 1250:69-86.

Voight, B., and S. Ferguson. 1988. Snow avalanches: a growing hazard to Americans. Ground Failure (National Research Council) 4:12-15.

Voight, B., H. Glicken, R. J. Janda, and P. M. Douglass. 1981. Catastrophic rockslide-avalanches of May 18. U.S. Geological Survey Professional Paper 1250:347-377.

Voight, B., and W. G. Pariseau. 1978. Rockslides and avalanches: an introduction. Pp. 1-67 in Rockslides and Avalanches, 1: Natural Phenomena. Amsterdam, Netherlands: Elsevier Scientific Publishing Co.

Voight, B., R. J. Janda, H. Glicken, and P. M. Douglass. 1983. Nature and mechanics of the Mount St. Helens rockslide-avalanche on 18 May 1980. Geotechnique 33:243-273.

Voitkovskiy, K. F. 1987. Snow cover stability on slopes and avalanche dynamics. International Association of Hydrological Sciences Publications 162:337-351.

Waitt, R. B. 1990. Swift snowmelt and floods (lahars) caused by the great pyroclastic surge at Mount St. Helens Volcano, Washington, 18 May 1980. Bulletin of Volcanology 52:138-157.

Waitt, R. B., T. C. Pierson, N. S. MacLeod, R. J. Janda, B. Voight, and R. T. Holcomb. 1983. Eruption-triggered avalanche, flood, and lahar at Mount St. Helens—effects of winter snowpack. Science 221(4618):1394-1396.

Watters, F. J., and K. Swanson. 1986. Sensor frequency, wave guide orientation and type, and their influence on acoustic emission monitoring of snow pack stability. Pp. 81-85 in Proceedings of International Snow Science Workshop, Tahoe, California.

Wells, K. 1987. Covering risk on the slopes. New York Times, January 4, XX-29.

Wilbour, C. R. 1986. The Chinook Pass avalanche control program. Pp. 195-200 in Proceedings of the International Snow Science Workshop, Tahoe, California.

Williams, H. 1934. Mount Shasta, California. Zeitchrift Vulkanologie 15(4):225-253.

Williams, K. 1975. The snowy torrents: avalanche accidents in the United States, 1967-1971. USDA/Forest Service General Technical Report RM-8. Fort Collins, Colorado: U.S. Department of Agriculture.

Williams, K. 1986. Colorado Avalanche Information Center Annual Report, 1985-1986. Denver, Colorado: Colorado Department of Natural Resources.

Williams, K., and B. Armstrong. 1984a. The Snowy Torrents: Avalanche Accidents in the United States, 1972-1979. Jackson, Wyoming: Teton Bookshop Publishing Co.

Williams, K., and B. Armstrong. 1984b. Colorado Avalanche Information Center Annual Report, 1983-1984. Denver, Colorado: Colorado Department of Natural Resources.

Witkind, I. J., P. A. Hoskins, V. L. Lindsey, and E. L. Mitchell. 1972. Map showing snow avalanche probabilities in the Henrys Lake quadrangle, Idaho, and Montana. U.S. Geological Survey Miscellaneous Investigations Series I-78-1. Scale 1:62,500.

Woodrow, R. E. 1986. Backcountry reports and ski-area boundary management guidelines. USDA/Forest Service, White River National Forest, August 1986 (distributed at National Avalanche School, Denver, Colorado, November 1986).

Yanlong, W., Zichu, X., and Z. Zhang. 1980. Prevention of avalanches in the Gunes Valley in Tienshan, China. Journal of Glaciology 26(94):520-521.

Zalikhanov, M., L. A. Akaeva, and Z. V. Vorokov. 1987. Avalanche distribution in the Causasus. International Association of Hydrological Sciences Publications 162:627-628.